AN INTRODUCTION TO
BEHAVIOR GENETICS

AN INTRODUCTION TO
BEHAVIOR GENETICS

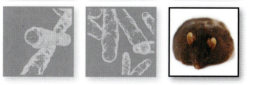

Terence J. Bazzett

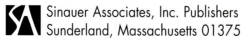

Sinauer Associates, Inc. Publishers
Sunderland, Massachusetts 01375

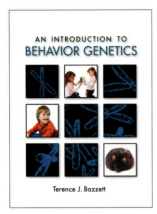

An Introduction to Behavior Genetics
Copyright 2008

For information address
Sinauer Associates,
23 Plumtree Road,
Sunderland, MA 01375 U.S.A.

FAX: 413-549-1118
Email: publish@sinauer.com
Internet: www.sinauer.com

Library of Congress Cataloging-in-Publication Data

Bazzett, Terence J., 1962-
  An introduction to behavior genetics / By Terence J. Bazzett.
      p. cm.
  Includes bibliographical references and index.
  ISBN 978-0-87893-049-4
  1. Behavior genetics. I. Title.
  QH457.B37 2008
  591.5—dc22
                                    2008027015

Printed in China
4 3 2 1

Dedicated to my parents

*In appreciation of the restless genes and
the nurturing environment*

CONTENTS IN BRIEF

TABLE OF CONTENTS

PART II
HERITABILITY, ENVIRONMENTAL INFLUENCES, AND METHODS OF STUDY   113

PART III
GENETIC INFLUENCES ON BEHAVIOR AND BEHAVIORAL DISORDERS 229

PART IV
OTHER GENETIC INFLUENCES, COUNSELING, AND THE FUTURE   349

PREFACE

Teaching Biological Psychology and Behavioral Pharmacology courses over the past two decades, I have noticed a growing trend in the literature. With each new edition of textbooks on these topics there is an increasing emphasis on the contributions of genetics to behavior. Not surprisingly, this increase has paralleled a rapidly expanding literature that has been stimulated by the Human Genome Project. In an attempt to stay progressive and to ensure that my students had the opportunity to learn about this highly relevant field, I designed a Behavior Genetics course for upper-level undergraduate students in the natural and social sciences. I have worked for several years developing this course, with the primary goal of educating students, and with a secondary objective of conveying the importance of this field and the excitement it is generating among researchers from many disciplines. In the process of teaching this course I have been delighted with the way my students have responded to the material being presented. Each class, consisting mostly of psychology and biology majors, has seemed to grasp the relevance of the subject matter and they have genuinely enjoyed the process of learning the associated concepts that are both wide ranging and thought provoking.

Although the content of this course is inherently captivating, one difficulty I encountered was finding an appropriate text to introduce this topic to students. Few textbooks have been published that focus specifically on Behavior Genetics, and of those, none seemed to adequately cover the material in a manner that students find accessible, interesting, and contemporary. The seeming void in the literature led me to compose a pedagogically sound text that would engage students as they worked toward learning basic concepts of this diverse and expansive field. A related goal was to develop a text that might inspire instructors to consider developing a new course that could be integrated into their curriculum.

Writing a textbook designed to introduce students to the field of Behavior Genetics proved to be highly challenging because the intersecting of scientific disciplines is so complex. Even with the assumption that students entering the course will have some background in the basic concepts of biology and chemistry, the range of general information needing explanation is formidable. This meant that selecting the material to be covered in each chapter required difficult choices about what information should be included and what information could be reasonably omitted. In the end, I was generally pleased with the breadth of material covered, which should appeal to a rather broad audience. This range of information also allows instructors to select and elaborate upon the topics they feel are most essential for meeting the learning objectives of their individual course.

It is my most sincere hope that this text intrigues readers as much as it educates them. My objectives for *An Introduction to Behavior Genetics* are to introduce readers to a field that expands beyond the enclosed pages and to provide a catalyst for students seeking to increase their knowledge of the biological factors that influence behavior. I also hope this text becomes a useful tool for instructors wishing to elaborate on its contents. If, after reading this text, some readers are inspired to pursue a more advanced education in the field of Behavior Genetics, than this book will have achieved an ultimate level of success.

Acknowledgments

I would like to extend my sincere gratitude to those individuals who inspired and guided me throughout the process of developing this textbook. From the outset Graig Donini offered constructive and insightful editorial suggestions, helping to give shape and structure to my intangible vision of this text. Chelsea Holabird provided constant input and feedback during the publication process, keeping me moving at an efficient pace that always seemed perfectly balanced with the workload. Laura Green was a keen fact checker whose suggestions helped strengthen the scientific content of this text. Norma Roche was a remarkable copyeditor, smoothing my inconsistent and sometimes cumbersome writing style into a text that reads beautifully from start to finish. David McIntyre constantly impressed me with his selection of seemingly perfect photos from vast collections, his efforts to track down unique photographs from around the world, and his willingness to contribute his own photographs to this book. His efforts created a visual experience for this text that exceeded my expectations, and working with him was both interesting and rewarding. Joanne Delphia created an elegant book design, including beautiful chapter opener pages, which were complemented by strong graphic images that she and David McIntyre chose. Joanne also created the dynamic and contemporary cover design. My thanks also to the illustrators at Dragonfly Studio who, at times, worked from my crudely scrawled representations to create illustrated figures capable of conveying complex concepts beyond what could be represented in words alone.

Laura A. Baker (University of Southern California), Gary L. Dunbar (Central Michigan University), Arthur P. Mange (University of Massachusetts, Amherst), and Martha A. Mann (University of Texas at Arlington) each earned my gratitude and my respect for their constructive critiques offered during various stages of the publication process. I would also like to thank Dr. Roger Albin, Dr. Jill Becker, and Dr. Elaine Hull, each of whom played a crucial role in fostering my interests in both teaching and research. From SUNY Geneseo I would like to thank Dr. Steven Kirsh and Dr. Margaret Matlin for their advice and encouragement, Dr. Wendy Pogozelski for her technical insight, and the SUNY Geneseo Research Foundation for their support. Finally, I offer my most sincere thanks to my wife Stacy who offered unconditional support and inspiration while enduring countless hours of conversation about this project over the past several years.

Instructor's Resource Library

The Instructor's Resource Library CD, available to qualified adopters of *An Introduction to Behavior Genetics*, includes all of the figures (including photos) and tables from the textbook in a variety of formats, making it easy to incorporate images from the book into your lecture presentations and other course materials. The Resource Library includes high-resolution and low-resolution JPEG files as well as ready-to-use PowerPoint® slides of all figures and tables. All images have been formatted and color-corrected for excellent image quality when projected in class.

PART I
AN INTRODUCTION TO BEHAVIOR GENETICS

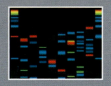

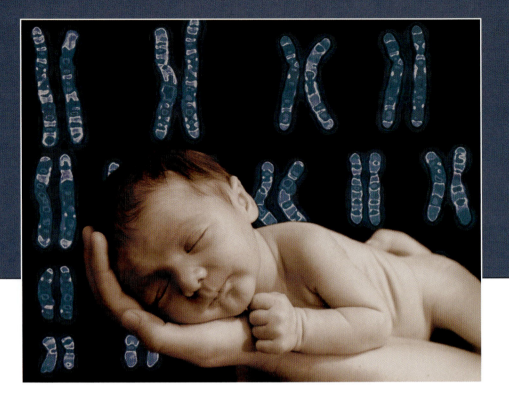

1

Introducing Behavior Genetics: Origins and History

Throughout history, philosophers and scientists have striven to understand the basis of human behavior. And while each successive generation has brought with it a seemingly clearer picture of why humans behave as they do, the base of information used in the explanation has grown to unwieldy proportions. Molecular discoveries, including the unraveling of DNA structure and function, represent some of the most significant recent contributions to behavioral research. Their significance lies in the reality that all previous discoveries suggesting organic influences on behavior (e.g., neurons, neurochemicals) necessitate a genetic basis for those influences. Furthermore, researchers have come to accept the theory that intangible factors (such as environmental influences and experiences) can directly affect genetic traits. In other words, genetics represents a focal point to which many diverse theories of behavior can be linked. This chapter describes some events in the evolution of the field of behavior genetics. Interestingly, although formal recognition of this field has come only within the past few decades, many of the basic concepts forming its theoretical foundations were put forth nearly a century ago. Indeed, it is an area of research that has accelerated rapidly as technological advances have become available. Furthermore, although behavior genetics is now formally recognized as a distinct area of research, essential data are being produced by researchers from a wide range of disciplines.

A truly integrative science, behavior genetics fosters highly productive collaborations and raises appreciation for the value of many research fields, all contributing to the single long-standing goal of describing the basis of variations in human behavior.

Defining Behavior Genetics

Behavior genetics is an intriguingly complex field of research. Considering that complexity, it seems appropriate to begin by defining some basic terminology. **Behavior** should be thought of not only as overt actions, but also as the intrinsic emotions, moods, and personality of an organism. **Genetics** can be broadly defined as both the physical nature of genes and the patterns of heredity that predict how those genes will be transmitted from one generation to the next. Thus, the field of **behavior genetics** is concerned with identifying genes and inheritance patterns underlying actions, emotions, moods, and personality.

Behavior genetics contributes to, and benefits from the contributions of, a number of basic and applied research fields (Table 1.1). The study of behavior is of primary interest to several fields, including psychology, sociology, and anthropology. Although some researchers from these fields also study aspects of heredity, the broader field of genetics is dominated by biologists, chemists, and geneticists. Behavior genetics research findings are also becoming more prevalent in textbooks describing evolution and basic neuroscience. These findings have had applied benefits as well, directly influencing practices in medicine, nursing, and psychiatry. Finally, the study and application of law and the shaping of government policy are being increasingly influenced by behavior genetics findings. As the field of behavior genetics grows, it will continue to attract the collaboration of scientists and practitioners from diverse areas, becoming a truly integrated field of research.

Effects of genetics on physical development

An in-depth understanding of genetics is not required to recognize that physical traits of offspring are strongly influenced by their parents. In fact, humans have bred domesticated animals for certain physical traits (i.e., size, color, hair length) since they first domesticated animals, thousands of years before **Charles Darwin** (1809–1882) first formally articulated the concept of heredity in his seminal work *The Origin of Species*, published in 1859. Shortly thereafter, in 1866, **Gregor Mendel** (1822–1884) proposed that physical features of pea plants, and presumably other organisms, are influenced in a highly predictable manner by inherited "factors," later to be known as genes.

TABLE 1.1   Fields of Research That Benefit from and Contribute to Behavior Genetics

---

Biology: The study of living organisms
  Human biology: The study of biology with a focus on humans
  Genetics: The study of genes and heredity
  Molecular biology: The study of molecules associated with cell structure and function

Psychology: The study of mental processes and behavior
  Biopsychology: The study of the biological basis of mental function and behavior
  Behavioral neuroscience: The study of brain structure, chemistry, and physiological processes as a basis for behavior
  Behavioral pharmacology: The study of chemical and pharmacological basis of mental function and behavior

Chemistry: The study of matter and its interaction with energy
  Organic chemistry: The study of the chemistry of carbon compounds
  Biochemistry: The study of the chemistry of living beings

Anthropology: The study of humans and nonhuman primates
  Biological anthropology (physical anthropology): The study of human evolution, adaptation, and variation

Sociology: The study of human societies and behavior
  Sociobiology: The study of biological factors that influence social behavior
  Evolutionary sociology: The study of evolutionary theory to better understand social behavior
  Sociology of deviance and criminology: The study of behaviors that violate social norms

Health professions: The delivery of preventive or responsive health care
  Genetic counseling: Educating individuals and guiding their decisions in matters related to hereditary factors in mental and physical health
  Gene therapy: Altering gene structure in an effort to reduce or eliminate genetically based dysfunction

---

Since the time of Darwin and Mendel, researchers have identified and come to understand many facets of genetic influences on physical features. In simple terms, genetic factors provide the instructions used to build proteins. Protein construction, in turn, creates the physical structure of the individual organism. Variations in genetic makeup from one individual to the next create unique protein structures that result in distinctive features (Figure 1.1). Darwin proposed that within any given species, a superior (i.e., stronger, healthier) organism is more likely to survive and reproduce successfully than is a developmentally in-

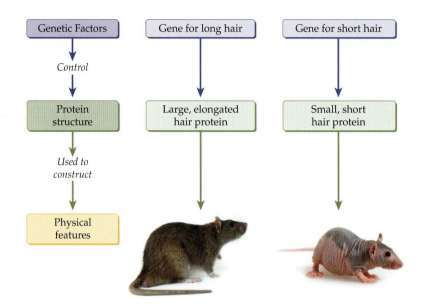

FIGURE 1.1 In a simple example of genetic effects on physical structure, variation in certain genetic codes may result in production of either large, elongated hair proteins or very short, fine hair proteins. Darwin suggested that when genetic factors produce traits that enhance survival, such as long hair in a cold climate, those traits will be passed on to offspring. Mendel's findings complemented Darwin's theories by showing a predictable mathematical basis for genetic transmission from one generation to the next.

ferior conspecific. Mendel showed that genetic factors, which influence physical features of an organism, are reliably passed on to offspring, resulting in physical features being shared between parents and offspring and between siblings. Together, these findings show the association between genetics, physical and behavioral characteristics, and the process of natural selection, creating the foundation for subsequent theories of heredity.

Effects of genetics on behavior

Although the effects of genetics on behavior are perhaps less intuitive than the effects of genetics on physical development, history shows a long-standing appreciation for the effects. For example, just as early selective breeding of domesticated animals was used to control physical traits, it can be assumed that **temperament** (enduring behavioral traits) was a consideration in the breeding of species such as dogs and horses. While Darwin's specific examples of heritable traits in *The Origin of Species* focused on physical attributes, it appears that he also recognized the potential heritability of behavioral features when composing the following passage in his manuscript's introduction:

> As many more individuals of each species are born than can possibly survive; and as, consequently, there is a frequently recurring struggle for existence, it follows that any being, if it vary however slightly in any manner profitable to itself, under the complex and sometimes varying conditions of life, will have a better chance of surviving, and thus be naturally selected. From the strong principle of inheritance, any selected variety will tend to propagate its new and modified form.

The general wording of this statement, particularly the reference to variation "in any manner profitable," suggests that Darwin did not wish to limit his theory to physical traits alone.

Recall that the definition of behavior given earlier in this chapter included overt actions as well as intrinsic features such as personality. It is now readily accepted that at least some features of behavior are highly heritable and thus closely mediated by genetics. Dogs offer an excellent example of this concept, as some breeds are characterized predominantly by behavioral traits, such as pointing, retrieving, or herding. In addition, some breeds have strong temperamental tendencies, such as being aggressive, passive, or playful. The ability to reliably produce these distinctive behavior and personality traits in certain breeds, within a single species, shows the power of heredity in establishing behaviors as well as physical attributes.

The explanation for how genetics creates physical structures is straightforward: Genes direct the construction of proteins. Proteins create the physical structure of the individual. Variations in genes create variations in physical features, and because genes are heritable, these variations may be passed on from parent to offspring. Now broaden this concept to include the understanding that behaviors are controlled by the brain, and that the brain contains billions of proteins that determine how it will function. Just as any two unrelated individuals look different because of variation in the proteins used to construct skin, bone, and hair, those same two individuals have significant variation in their brain proteins. This variation results in behavioral and personality traits that are as different as their physical features. In short, it is the genetically directed physical makeup of the brain that underlies much of the behavioral variation seen between any two individuals of the same species. In this regard, there is an assumption that behaviors cannot be separated from the physical features of the brain controlling those behaviors (Figure 1.2).

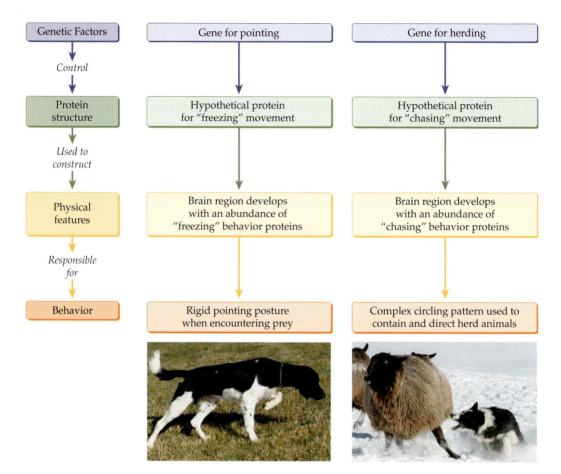

FIGURE 1.2 In a simple hypothetical example of genetic effects on behavior, the same sequence of events that produces genetic effects on physical structure is followed (see Figure 1.1). In this case, however, the physical structure of interest is a brain protein that directly influences behavior. Genetic alterations in brain proteins can create a wide range of behaviors.

The influence of environment on physical development and behavior

Although genes direct construction of the proteins that form the basis for physical development and behaviors, protein viability is also influenced by external factors. The most basic example of such an influence is an organism's need for environmental resources to provide the energy and chemical building blocks for manufacturing proteins. In animals, food consumption is a good example of this interplay between genetics and environment. It is easy to imagine the neg-

(A)

(B)

FIGURE 1.3 (A) To ensure their survival, the genes of wolves produce protein structures in their brains that promote aggressive behavior. (B) To function effectively in a hierarchical social group, however, some wolves must suppress aggression toward others within the pack. In this case, behavioral tendencies directed by genetics are significantly altered by environmental factors. It is assumed that such environmental factors physically alter protein structures in some brain regions to ensure enduring behavior patterns.

ative effects of prolonged food deprivation (starvation) on both physical development and behavior, even when the genetic instructions are flawless.

A more subtle example of environmental influences on physical structure and behavior would be the effects of social interactions on brain development. Take, for example, animals living in groups that form social hierarchies based on dominance. When an animal's displays of aggression are met with submission by others, the animal learns that aggressive behaviors place it at the top of the social hierarchy. Conversely, an animal whose aggressive displays are met with even more aggressive counterattacks learns to display submissive behaviors associated with the bottom of the social hierarchy (**Figure 1.3**). The "learning" of aggressive and submissive behaviors is, in fact, a change in protein function or structure within the animal's brain. In this example, genetics may be thought of as driving the development of brain proteins that mediate social behaviors. But it is ultimately environmental influences (social interactions) that fine-tune brain proteins to determine the specific form those social behaviors will take (aggressive or submissive).

It is also worth noting that there is the potential for a unique interplay between genetic and environmental effects on behavior. Just as environmental factors can influence genetically controlled behaviors, genetically controlled factors can influence an individual's environment. **Down syndrome** is the effect of the addition of an extra chromosome 21 to the human genome. Although this syndrome is often associated with impaired intellectual ability, it also re-

FIGURE 1.4 Most children born with Down syndrome exhibit characteristic physical features that include low-set ears, upward-slanted eyes, and a small mouth with a protruding tongue. Adults who identify these features in a child tend to assume that the child has intellectual impairments and to interact with the child in a remedial manner. Such interaction has the potential to slow the child's intellectual development, regardless of genetic influences on brain function.

sults in several characteristic physical features (Figure 1.4). Individuals with these physical features are frequently viewed as mentally disabled and, as a result, may be subjected to lower expectations for intellectual achievement. In this case, it is reasonable to conclude that the intellectual impairment of some of these individuals could result in part from environmental factors. Furthermore, these environmental factors (the lowered expectations of others) are a direct response to genetically determined physical characteristics.

Less obvious but more prominent examples of such genetically influenced environmental factors are abundant. Being male (a genetic trait), for example, is typically associated with aggression. During human development and socialization, others tend to tolerate, and in some cases expect, greater levels of physical assertion from males than from females. Similarly, genetic contributions to anxiety or shyness could result in lower levels of social contact. Inadequate social interaction could then, in turn, lead to impaired development of social skills, further exacerbating anxiety or shyness. Clearly, there is a potential for many interactions between genetic factors, physical structure, environment, and behavior.

Thus, when defining and describing behavior genetics research, a large net must be cast. First, the effects of genetics on physical development must be considered. The physical development of brain structures, in turn, influences all aspects of behavior. The ultimate form and function of brain structures, however, is not a result of genetics alone, but may also be influenced by a wide range of environmental factors. For all of these reasons, behavior genetics is a complex field of study, and that complexity makes the seemingly simple task of defining this field a formidable undertaking.

Origins of Behavior Genetics

Though recent highly publicized events, such as pilot studies in human gene therapy and the completion of the Human Genome Project, may give the impression that behavior genetics is a new field of research, this is not the case. In fact, behavior genetics has a rather colorful history spanning nearly two centuries. During this time, the field has drawn heavily from a diverse range of research areas, in a manner similar to that seen in the evolution of the field of behavioral pharmacology. Presented here is a summary of noteworthy events in the history of behavior genetics, as well as a description of some parallels with behavioral pharmacology.

A brief history

To appreciate how behavior genetics has come into its own as a distinct field of research, a brief look at some key historical events may be useful (Table 1.2). As mentioned previously, Mendel and Darwin both contributed information

TABLE 1.2 A Chronology of Events in the Development of the Field of Behavior Genetics

YEAR	EVENT
1850	Darwin: *The Origin of Species* (1859) Mendel: First article on heritability (1866) Galton: *Hereditary Genius* (1869)
1870	↑ Lack of technology impeded practical development of early concepts ↓
1940	World War II (1939–1945): Nazi policies evolve from skewed interpretation of eugenics
1950	DNA photographed; researchers theorize a simple double-helical structure
1960	Identification of coding system used for 20 amino acids
1970	Advancement in techniques to visualize and manipulate DNA
1980	Identification of Huntington disease gene; start of the Human Genome Project
1990	Completion of the Human Genome Project; first gene therapy trials Publication of *The Bell Curve* and *The g Factor* raises new ethical concerns
2000	Identification of genes for many aspects of physical development, as well as behavioral disorders and normal behaviors

FIGURE 1.5 Sir Francis Galton (1822–1911), cousin to Charles Darwin, was a mathematician with an intense interest in how complex behaviors are influenced by heredity. Galton was the first to formally propose the concept of eugenics as a means of enhancing the overall health and prosperity of the human race.

that was vital to forming concepts of heredity in the late 1800s. At about the same time, **Francis Galton** (1822–1911), a mathematician and cousin of Darwin, became interested in the role heredity played in many facets of development (Figure 1.5). In his 1869 publication *Hereditary Genius: An Inquiry into Its Laws and Consequences*, Galton argued that, much like physical traits, some features of complex behavioral traits, such as intelligence, were inherited. His insights into the potential for genetic influences on such traits earned him the title "Father of Behavior Genetics" from some historical observers. In his writings, Galton suggested that the health and viability of the human race could be improved if human breeding patterns were more closely observed and controlled, a theory he termed **eugenics**.

Although Galton's theories were highly insightful, his work had unanticipated negative consequences. For example, though his theory of eugenics was conceptually sound, its application would have required unreasonable restrictions on social behaviors and thus had the potential to give rise to serious human rights violations. For example, some early advocates of eugenics argued that sterilization of mentally inferior individuals could be justified, as the consequences would ultimately benefit future generations and society as a whole. In fact, one prominent advocate of eugenics was **Adolf Hitler** (1889–1945), leader of the Nazi regime of Germany, who molded the basic concepts of this theory into a pathological quest to generate a "master race" through a process of ethnic, racial, and religious genocide (Figure 1.6). In the aftermath of World War II, concerns arose regarding future research into hereditary aspects of complex human behaviors. Interest in behavior genetics research waned for the next several decades presumably due in part to concerns over its association with eugenics and Nazi policies. In fact, the skewed interpretation of eugenic theory formulated by the Nazi regime is still frequently cited in contemporary arguments against genetic engineering and gene therapy. Box 1.1 describes some recent controversies raised by interpretations and applications of behavior genetic research.

While reaction to Nazi Germany's policies may have repressed some efforts to explore the effects of specific genetic processes on human behavior, basic genetic research continued to flourish into the mid-1900s. In fact, some of the most significant advances in the field of genetics occurred in the 1950s, when the first high-resolution X-ray images of DNA were produced, allowing researchers to evaluate its physical structure (see Box 2.1). Shortly after those images were produced, researchers proposed that DNA was a relatively simple arrangement of bases bonded in a double-helical configuration. This understanding of the basic structure of DNA allowed scientists to further delineate individual components of both DNA and RNA in the 1960s, leading to the discovery of the coding system used to produce each of the 20 amino acids found in humans.

With this knowledge of the basic components of DNA and how they are arranged to form code-containing molecules, researchers in the 1970s developed techniques not only to visualize DNA, but also to manipulate the DNA molecule in an effort to alter its function. It was also during this decade that the idea of commercial applications of **genetic engineering**, the manipulation of genes to produce desirable traits, generated widespread interest. The concept of commercial genetic engineering created

FIGURE 1.6 Adolf Hitler (1889–1945) took an extreme and violent approach to applying Galton's theory of eugenics by directing a large-scale program of systematic genocide, based on religion, race, and ethnic background. Hitler's skewed interpretation and application of eugenics raised valid moral and ethical concerns that are still voiced today in debates about how genetic research should be conducted and applied.

BOX 1.1   Behavior Genetics: Apply with Caution

Few other areas of research conjure the degree of combined excitement and angst that the study of behavior genetics does. Considering the nature of the information presented in this text, and its unique potential for misinterpretation and misuse, an early word of caution seems appropriate. Decoding the physical essence of behavior is an awesome responsibility that requires a great deal of restraint in interpreting and applying research findings. The example of Adolf Hitler's use of the concepts of eugenics as a rationale for mass genocide is one notable, albeit extreme, case of misuse of the theories of behavior genetics.

More recently, controversy arose when Richard J. Herrnstein and Charles Murray presented their interpretation of data on heritable aspects of intelligence in their 1994 best seller *The Bell Curve*. These authors suggested that IQ differences between ethnic groups could be explained largely by genetic factors (a claim reiterated by Arthur Jensen in his 1998 publication *The g Factor: The Science of Mental Ability*). *The Bell Curve* further proposed that genetically mediated low intelligence was a root cause of crime, welfare dependence, and single parenting. While stopping short of suggesting a revival of eugenics, Herrnstein and Murray did

(Continued on next page)

BOX 1.1 Behavior Genetics: Apply with Caution (*continued*)

propose eliminating welfare programs, which they believed subsidized the birth of low-intelligence individuals, and developing programs to encourage women from higher socioeconomic classes to have more children ("Genome research," 1995; "Statement," 1996).

This misrepresentation of behavior genetics data caused great concern among researchers, who then worked through the National Institutes of Health and the Department of Energy to form the Joint Working Group on the Ethical, Legal and Social Implications (ELSI) of Human Genome Research. Largely in response to public outcry over Herrnstein and Murray's assertions, the ELSI Working Group published a statement condemning the misuse of behavior genetics research to inform social policy ("Statement," 1996).

In 2005, at the National Bureau of Economic Research Conference on Diversifying the Science and Engineering Workforce, Dr. Lawrence Summers, then president of Harvard University, postulated several reasons why women had failed to make significant gains in securing faculty positions in science and engineering departments at the top research institutions. In addi-

tion to citing discrimination and socialization as possible factors, Summers offered his interpretation of recent behavior genetics research that had shown a tendency for the sexes to differ in some cognitive functions. The controversy that surrounded Summers' statement focused on concern that he had interpreted research findings as evidence of male genetic superiority in several areas of research. Summers subsequently apologized for publicly voicing his comments, which he described as "explicitly speculative." By that time, however, the damage was viewed by many as irreversible. In 2006, Summers resigned his position as president of Harvard, a move presumably motivated in part by his controversial interpretation of behavior genetics research.

It is likely that behavior genetics will be at the heart of many more controversies in the future. With this in mind, consider carefully how the information summarized in this text will be interpreted and disseminated. Behavior genetics is indeed an area with great potential, both for scientific discovery and for inciting social controversy.

concerns that such ventures should be subject to some form of regulatory oversight. As a result, scientists began formal consideration of limitations on their work based on moral and ethical factors.

Technological advances allowed researchers in the 1980s to identify specific genes associated with heritable diseases. One particular catalyst for behavior genetics research was the discovery of a gene for **Huntington disease**, a neurodegenerative disorder that produces significant motor disturbances as well as psychological decline. (Readers who notice that the name "Huntington disease" does not have a possessive form will find an explanation for its absence in Box 1.2.) By the end of the 1980s, planning had begun for the **Human Genome Project**, a

BOX 1.2   Names of Diseases: Possessive or Not Possessive?

Possessive or not possessive … that is the question. It is sometimes confusing to look through the literature and see that names of diseases may or may not contain an apostrophe and the letter s. For example, a literature search for "Huntington disease" may generate different results from an identical search for "Huntington's disease." So which name is correct?

Technically, "'s" indicates a possessive noun form. Thus, when "'s" is added to a disease name (i.e., Huntington's), it denotes a disease belonging to the person with that name (or more correctly, that the person with that name contracted the disease). Dr. George Huntington is generally credited with publishing the first paper describing the disease that bears his name, but he did not suffer from the disease. Therefore, the correct denotation is "Huntington disease." On the other hand, Lou Gehrig, a professional baseball player in the 1930s, was diagnosed in the prime of his career with amyotrophic lateral sclerosis (ALS). ALS is often referred to as "Lou Gehrig's disease." In this case, the "'s" is correctly used.

To put this rule in perspective, the Lincoln Memorial in Washington, D.C., is not "Lincoln's Memorial." Chicago baseball fans would cringe if they heard "Comiskey's Park" or "Wrigley's Field." The names of both ballparks commemorate individuals, just as the memorial in Washington is dedicated to Lincoln (not owned by Lincoln). Similarly, many buildings on college campuses are named for individuals. In these cases, "'s" is rarely used.

(A) George Huntington, a physician, was the first person to formally describe a particular inherited disease characterized by uncontrolled movements and cognitive decline. To acknowledge his contributions, the disorder was later named "Huntington disease" in his honor. (B) Lou Gehrig was a popular and gifted professional baseball player who was diagnosed with amyotrophic lateral sclerosis (ALS) in the prime of his athletic career. Because of his high profile, the disease he contracted became associated with his name and is now commonly referred to as "Lou Gehrig's disease."

Usage of "'s" may seem to be a trivial point, considering that the majority of researchers and teachers are guilty of using it incorrectly at some time. It is also the case that popular usage can sometimes overshadow correct usage. In fact, most organizations dedicated to providing information on diseases use the incorrect (though more popular) possessive forms of their names (Huntington's Disease Society of America, Alzheimer's Disease Education and Referral Center, etc.). However, knowledge truly is a valuable tool, and even this small bit of information may be useful for students or researchers wishing to focus or broaden their literature searches.

historic collaboration of laboratories and scientists with the common goal of identifying every human gene. This project not only accomplished all of its primary research objectives, but reached those goals well ahead of the projected 15-year timeline.

Through the 1990s and into the first decade of the twenty-first century, behavior genetics researchers have been applying the vast amounts of information accumulating as a result of surging technological and research discoveries. **Gene therapy**, the manipulation of genes to reverse the effects of diseases, is now in early stages of human trials. There is also a growing sense in the scientific community that behavior genetics has begun the shift from a largely basic field of research to a more applied field. In particular, hope is rising that psychiatric maladies currently treated with chronic drug administration may one day be reversed with a single application of gene therapy.

Relationship to behavioral pharmacology

If gene therapy seems futuristic, it might be worth noting that less than a century ago, pharmacological treatments for psychiatric disorders were largely nonexistent. In fact, the history of behavior genetics has many parallels in the history of **behavioral pharmacology**, the field concerned with the effects of drugs on behavior. It was in the early 1900s that researchers began systematically investigating and measuring the effects of psychoactive drugs on behavior. At that time, I. V. Zavadskii, a colleague of Ivan Pavlov, was among the first to assess the effects of alcohol, cocaine, morphine, and caffeine on conditioned salivating behavior (Laties, 1979). About 30 years later, renowned behaviorist B. F. Skinner and his colleague W. T. Heron published the first report of psychoactive drug effects on conditioned operant behavior patterns (Skinner and Heron, 1937). But most researchers agree that the field of behavioral pharmacology came into its own when a series of studies in the 1950s showed that behavior patterns previously found to be maintained by natural positive reinforcers (e.g., food, water) could be readily modified by drug administration (Barrett, 2002). This methodological finding opened the door for a wide range of tests in which researchers analyzed drug effects on normal behavior patterns in animals. From this research, hypotheses about the chemical basis of specific behaviors were formed, allowing scientists to develop, assess, and modify drug treatments for behavioral abnormalities. The 1950s and 1960s were marked by the discovery of a wide array of pharmacological treatments for previously untreatable psychiatric maladies, including schizophrenia, depression, and anxiety disorders. Careful analyses of those drugs over the next several decades led to further refinement of established psychopharmacological treatments as well as the discovery of novel treatments for many additional psychiatric disorders.

Throughout those years, researchers gained invaluable insights into the chemical basis of both normal and abnormal brain function. To realize these many accomplishments, researchers from a wide range of scientific fields collaborated in generating a cohesive understanding of pharmacology, biology, anatomy, chemistry, and behavior. In time, from these many diverse fields of research, the unique and influential independent field of behavioral pharmacology evolved.

Throughout this text, the interplay between behavioral pharmacology and behavior genetics will be revisited. It will become clear that some findings from behavioral pharmacology offer invaluable insights into genetic influences on behavior. Likewise, findings from behavior genetics have the potential to enhance applications of behavioral pharmacology by allowing its practitioners to maximize the effectiveness of drug treatments and minimize their side effects. Indeed, in the future, it is likely that applications of these two fields will be increasingly complementary.

What to Expect from This Text

With the understanding of, and appreciation for, the origins of behavior genetics readers have gained in the previous section, they should have high expectations for a textbook on this topic. This text takes a broad perspective on the field of behavior genetics, and it assumes that readers have only a basic knowledge of specific topics in the field. Some background information, however, will be highly valuable when reading the subsequent chapters. Furthermore, readers should realize that many of the topics presented in this text parallel information presented in writings from other related disciplines. After outlining the kinds of background information that may be useful to readers, this section describes the organization of this text and the logic behind the sequence in which information is presented. Readers may find it helpful to use this preview to establish desired learning objectives before reading the remainder of the text.

Useful background knowledge

This book was developed as an introductory text with an assumption that its readers will possess a basic understanding of general psychology and human biology. It also presents some basic concepts from the fields of chemistry and physics. If scientific theories or concepts from any of these areas are not clearly understood, supplemental readings from introductory college-level textbooks in the areas of concern may be useful.

Laboratory methods are described in several chapters outlining processes of discovery in the field of behavior genetics. Readers who have completed a

FIGURE 1.7 The Human Genome Project Information Web site provides many excellent contemporary readings that supplement this text.

college-level course in research methods, or who have directly participated in laboratory research, will be best prepared to understand the methodology presented throughout this text.

Comprehension of most data from human heredity studies presented in this text will require a general understanding of statistical methods, particularly in correlation research. Readers unfamiliar with such statistical analyses would be advised to supplement this text with additional readings from a college-level behavioral statistics textbook.

Finally, it is highly recommended that readers visit the government-sponsored Web site developed in collaboration with researchers from the Human Genome Project (www.ornl.gov/sci/techresources/Human_Genome/home. shtml) before moving beyond this introductory chapter. This Web site contains a wide range of information and resource links directly related to many of the most current topics presented in this text (Figure 1.7).

Chapter content and organization

This chapter has outlined the origins of the field of behavior genetics and summarized some of its history. The chapters that follow have been organized in a manner designed first to create an understanding of the basic concepts of the field, and then to introduce more complex concepts built on that understanding. **Chapter 2** begins by outlining the physical nature of, and the relationship between, DNA and RNA. It then describes the structure of chromosomes and the function of genes. The ability to conceptualize protein construction as a primary function of specific genes is a crucial learning objective for this chapter.

Chapter 3 links this understanding of gene-mediated protein production to behavior by describing basic features of neuronal structure and function. Most people realize that behaviors are controlled by the nervous system, but behavior genetics researchers are particularly interested in the relationship between genes and neuronal function. In addition to describing how neurons function, this chapter notes some ways in which genetic variation can lead to alterations in normal brain activity. This concept of individual genetic differences producing a range of variation in brain structure and behavior carries over into many later chapters.

Chapter 4 outlines some of the basic methods used in behavior genetics research, emphasizing methods for visualizing DNA, RNA, and genes. The subsequent chapters on heredity draw heavily on the idea that genes can be "identified." This chapter offers some general insights into how genes and gene defects are located and how DNA "fingerprinting" is conducted. Finally, it reviews several prominent findings from the Human Genome Project. This project, arguably one of the greatest research undertakings of a generation, has the potential to influence many aspects of genetic research and health care. References to the Human Genome Project are abundant in subsequent chapters.

Whereas Part I focuses largely on research at the molecular level, Part II emphasizes the study of heritability and the effects of environment on behavior. **Chapter 5** summarizes the concepts of simple inheritance, built largely on the contributions of Gregor Mendel's seminal work on pea plants. But Mendel's original observations, though astute, were greatly oversimplified. As a result, many of his laws of inheritance have been amended or expanded over time to give researchers a more complete understanding of inheritance patterns. **Chapter 6** expands the simple observation that a single gene can produce phenotypic effects to the more complex idea that multiple genes may act together to influence a trait. Such polygenic traits constitute the major area of interest for behavior genetics researchers.

Drawing on the understanding of genes and heredity established in the first six chapters, **Chapter 7** introduces the concept of environmental influences on genetic effects. This chapter begins by reviewing the effects of evolution on physical structure and function, particularly through the process of natural selection. Whereas most readers will be familiar with the effects of some environmental factors on behavior, this chapter takes note of some less obvious but equally important factors. For example, it assesses the effects of the prenatal environment, and it describes environmental factors that may directly alter gene structure as well as gene function. It also summarizes current perspectives on the "nature versus nurture debate": the long-standing controversy over whether genetics or environment has a greater influence on the development of an individual. Just as Chapter 4 described the basic methods of researching molecular aspects of behavior genetics, **Chapter 8** summarizes the methods used to study heritability. This chapter describes the procedures used for trait breeding as well as those used for creating and testing genetically altered animal models. Human research using twin and adoption studies is also described, with an emphasis on how traits are followed in family histories.

The information in Parts I and II of this text provides a substantive knowledge base for understanding genetics and heredity. In essence, this first half of the text is focused on the "genetics" portion of behavior genetics. Part III shifts that focus to the "behavior" portion. Although readers interested in behavior are often most interested in learning how genetics contributes to abnormal behavior, **Chapter 9** begins by describing the influence of genes on the development of normal behavior patterns. This chapter describes the underlying effects of genes on the normal range of intelligence, mood, personality, and sexual behavior. Once the reader has a clear understanding of how genes are intended to work on these functions, the abnormal patterns of development described in Chapters 10–12 can be more fully appreciated as variations from these normal developmental patterns.

Chapter 10 surveys some influences of genetics on cognitive impairment. Although mental retardation is an obvious form of cognitive dysfunction for discussion in this context, learning disabilities and dementia represent two additional important classes of intellectual impairment. All of these categories of cognitive impairment have some genetic influences that are currently being studied in both humans and animal models. **Chapter 11** builds on the research presented in Chapter 10 to summarize findings on other psychopathologies in which cognitive dysfunction is a significant component. Schizophrenia and autism represent two primary examples. In addition, recent findings have shown the potential for genetic influences on the eating disorders anorexia and bulimia, both of which have a cognitive component. Finally, genetic contributions to addictive behaviors, which are frequently treated with cognitive-behavioral therapy, are covered in this chapter. **Chapter 12** surveys the influences of genetics on mood and anxiety disorders. Like cognitive disorders, these disorders are studied in both human and animal models. This chapter also explores the possibility of genetic contributions to antisocial personality disorder and criminal behavior.

Part IV moves on from the overview of behaviors and behavioral disorders commonly associated with this field. The chapters in this final part focus on related health issues, and the application of research findings to educating the public and to improving health services. **Chapter 13** focuses on genetic factors related to obesity and cardiovascular function. This chapter also highlights disorders directly associated with sex chromosome abnormalities. Some of these disorders have specific effects on sexual development, while others have more general effects on physical and psychological development. **Chapter 14** summarizes some advances in knowledge and technology that are currently being used to reduce or eliminate problems associated with abnormal gene function. The concepts of genetic counseling and gene therapy, briefly mentioned throughout the text, are presented in more detail in this chapter. In addition, the potential use of genetics research to more accurately tailor drug therapy, a concept known as pharmacogenomics, is discussed in this chapter.

The text concludes with a review of current perspectives and technologies related to behavior genetics, presented in **Chapter 15**. Of particular interest in this chapter is a glimpse into the future of formal education within the field of behavior genetics. Only recently recognized as an independent field of research, behavior genetics is fast becoming a focus of interest for established scientists, as well as students seeking a progressive and challenging career in research. In fact, this summary chapter ends very much where the text begins, with an acknowledgement that behavior genetics is an intriguingly complex field of research. A primary objective of this text, from Chapter 1 through Chapter 15, is to convey the excitement and potential of behavior genetics. Indeed, enthusiasm for research in this field has fueled a stunningly productive beginning. If this text can energize a mere handful of individuals to continue this pursuit, then the fundamental goal of the author will have been realized.

Chapter Summary

Behavior genetics is a complex field that attempts to elucidate both gene functions and heredity patterns as they relate to overt actions, intrinsic emotions, moods, and personality. Findings from behavior genetics research have advanced a broad spectrum of scientific fields and medical applications. They have also been used recently to establish legal precedents and to shape government policies.

Most people readily comprehend the directive influence of genetics on physical development and the fact that genes (and their associated physical traits) are often predictably passed on from one generation to the next. Sometimes less intuitive is the fact that, like overt physical traits, some distinct behavior patterns are also mediated by genes and consequently passed on from parent to offspring. The development of genetically influenced physical and behavioral traits can be conceptualized as a single process. More specifically, genes influence the physical structure of brain proteins, which in turn control behavioral traits. For many individuals, behavior genetics research may seem more tangible when behaviors are viewed as a by-product of the physical features of the brain.

It would be an oversimplification, however, to think of behavior as being strictly a result of brain activity. Events in the environment have the potential to dramatically alter gene-driven physical development. Interactions with the environment routinely alter normal brain function, and in some cases may change the physical structure of the brain, leading to long-term changes in behavior. The interplay between a genetically influenced brain structure and the wide array of environmental factors to which the brain is exposed results in limitless possibilities for behavioral development and expression. This means, in short, that the task of studying behavior genetics is formidable, to say the least.

The study of behavior genetics has diverse roots. Most students are familiar with Mendel's seminal work on hereditary patterns of single-gene traits and with Darwin's study of natural selection. Behavior genetics draws heavily from the theories of both of these early scientists. However, Sir Francis Galton is credited by many as being the first true scientist in the field of behavior genetics. In his book *Hereditary Genius*, he proposed that complex behavioral traits could be transmitted from one generation to the next in a manner similar to that proposed by Mendel for physical traits. By integrating this idea with the concepts of natural selection proposed by Darwin, Galton conceived the theory of eugenics. This theory proposes that the health of the human species could benefit from controlled breeding. Although the theory of eugenics is conceptually sound, its application could lead to major human rights violations (such as the sterilization of individuals with "undesirable" traits). In fact, Adolf Hitler's quest to create a "master race" by exterminating certain groups of individuals was rationalized in large part using eugenic theory.

The field of behavior genetics began to flourish in the 1950s with the advancement of basic scientific methods, including techniques that allowed researchers to visualize DNA and determine its structure. In the 1960s, the DNA coding system used to produce each of the 20 human amino acids was delineated. Researchers in the 1970s experimented with manipulating these codes in an effort to alter gene functions. Realization of the potential for such genetic code manipulation spawned the genetic engineering era. Further development of basic research techniques into the 1980s enabled researchers to identify individual genes involved in a wide range of genetic disorders, including the gene for Huntington disease, setting a benchmark for advancing research on a wide range of genetic disorders. Perhaps the most significant accomplishment in the history of behavior genetics was the completion of the Human Genome Project, which identified all genes in the human genome, creating a database of invaluable information that is now being shared by scientists around the world. One major application of all this information is gene therapy, the manipulation of genes to reverse the effects of disease. Although gene therapy is still highly experimental, the results of initial clinical trials point toward its incredible future potential for treating, and in some cases eradicating, many serious and fatal diseases.

The evolution of behavior genetics as a research field in some ways parallels that of behavioral pharmacology. Just as behavior genetics researchers are beginning to reveal the genetic basis of many complex behaviors, researchers from the field of behavioral pharmacology are uncovering the neurochemical underpinnings of those behaviors. Both fields have made significant strides from basic to applied research in the last several decades, and both fields share the goal of effectively treating previously untreatable neurological diseases and psychological disorders. In addition, the two fields have a strong reciprocal relationship, whereby advances in one field frequently benefit research in the other. This chapter concludes with an outline of the text, designed to preview the upcoming information and outline the logical sequence in which it is presented.

Review Questions and Exercises

- In the context of the field of behavior genetics, give a general definition of both "behavior" and "genetics."

- Describe how the early theories of Gregor Mendel and Charles Darwin provided a foundation for contemporary research in behavior genetics.

- Describe briefly how genetic influences on development of physical structures directly affect behaviors.

- Give a brief example of how genetic factors can alter an individual's environment. Explain why such alterations are important variables in assessing behavior.

- Explain why Sir Francis Galton is sometimes referred to as "The Father of Behavior Genetics."

- Describe how Adolf Hitler is associated with the history of behavior genetics.

- List several parallels between the fields of behavior genetics and behavioral pharmacology.

- Give an example of how the misinterpretation or misuse of behavior genetics research can have negative consequences for an individual or a society.

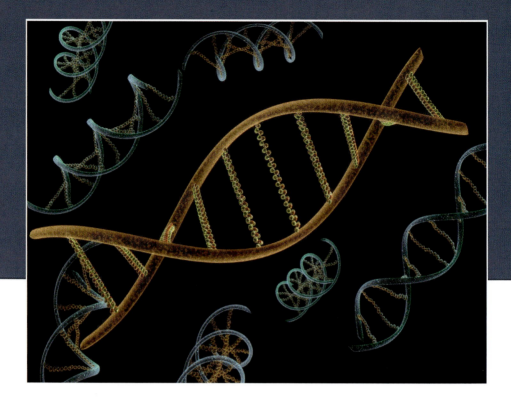

2

The Basis of Genetics: DNA, RNA, Chromosomes, and Genes

It seems that most people who choose to study behavior genetics are inherently more interested in behaviors than in the molecular basis of genetics. These people may think of reading and trying to comprehend the material presented in this chapter as a necessary evil for the greater goal of understanding this field. However, most readers are likely to find that they not only comprehend this material better than they expected, but also retain the information quite clearly. There are at least two explanations for this phenomenon. First, the structure and function of DNA and RNA are relatively straightforward. When compared with the complexities and variations seen in behavioral research, there is a simple elegance to these molecular structures. Second, many readers are likely to find the material far more interesting than they originally imagined. Although the analysis of molecular structures may seem complex and even slightly intimidating to those unfamiliar with the subject matter, always keep in mind that this information is the foundation for understanding behaviors. To reference an old adage, remember not to lose sight of the forest for the trees. As you read through this chapter and the rest of Part I, always keep in mind that the primary goal of learning these essential details is to better answer the much larger questions of interest. In the process, try to appreciate the many other fields of research that have contributed to the large and expanding science of behavior genetics.

DNA, RNA, chromosomes, and genes are the basic organizing principles of life. It is often said that "DNA is the blueprint for life." In fact, DNA does contain a wealth of information that affects both the structure and the function of an organism. But the influence of DNA cannot be adequately conceptualized without an understanding of its relationship to RNA or its organization into chromosomes and genes. To fully appreciate how genetics influences behavior, one must establish, at the very least, a working knowledge of each of these fundamental components. This chapter introduces some of these basic concepts. It is not designed as a comprehensive assessment, but rather as a summary of information that should enhance the reader's understanding of the material presented in subsequent chapters.

Before beginning this overview of the molecular components that underlie the field of behavior genetics, consider for a moment what researchers believe to be the central dogma of this field. DNA constitutes the most basic organized component of behavior. DNA dictates the assembly of RNA, which provides the basis for the construction of all proteins. These proteins, in turn, represent the physical basis for behavior. The brain cells and brain chemicals required for all actions are products of this relatively simple sequence of events. While reading through Part I of this text, do not lose sight of this important principle: The relationship between genetics and behavior has a definitive basic organization that follows a predictable and logical sequence of steps.

DNA

DNA is the abbreviation used for **deoxyribonucleic acid**, a molecule that can be broken down into several simple components. A *nucleic acid* is a large molecule composed of numerous subunits called **nucleotides**. Each of these subunits contains a five-carbon sugar, a nitrogenous **base**, and a phosphate group. *Ribose* is the five-carbon sugar in these DNA subunits. The *deoxy-* prefix denotes a missing oxygen atom in the sugar (Figure 2.1).

The four nitrogenous bases that make up nucleotides are **adenine**, **cytosine**, **guanine**, and **thymine**. Adenine and guanine each have a two-ring structure composed of four nitrogen atoms bonded with carbons. Molecules with this particular structure are called **purines**. Cytosine and thymine have an even simpler one-ring structure composed of two nitrogen atoms bonded with carbons (see Figure 2.1). Molecules with this structure are called **pyrimidines**.

The third component of a nucleotide is a phosphate group. Phosphorus is a simple mineral element found in many food sources (see Figure 2.1). In addition to its role in DNA binding, it is required for plant and animal growth and is a fundamental element in metabolic reactions. When nucleotides come together to form a DNA molecule, the phosphate groups bind the sugar molecules together.

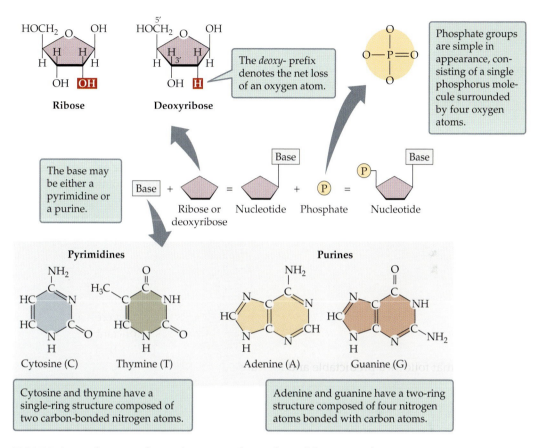

FIGURE 2.1 Ribose is a five-carbon sugar that is derived from six-carbon sugars in the diet (such as glucose). Ribose is converted to deoxyribose by replacement of one of the hydroxyl groups (OH) with hydrogen. This conversion results in the net loss of an oxygen atom. The four bases that make up DNA nucleotides are adenine, cytosine, guanine, and thymine.

Thus, DNA (as the name implies) is a large molecule consisting of sugars linked to bases and bound together by phosphates. This explanation is as simple as it seems. DNA has, in fact, a relatively simple chemical structure.

Structure

A strong chemistry background is not required to make sense of DNA structure. In simplified terms, DNA derives its structure from the bonding of molecules by a process called dehydration synthesis, whereby a water molecule is displaced, electrons are rearranged, and a covalent bond is formed. Through dehydration synthesis, any base can bond to any deoxyribose molecule to form a nucleotide (Figure 2.2A). Through the same dehydration synthesis process, phosphate groups bond to deoxyribose molecules to form the **DNA backbone** (Figure 2.2B). The result is a long chain of nucleotides, each of which contains one of the four bases. A single strand of nucleic acid is also labeled as 3′ (**three prime**) on one end and 5′ (**five prime)** on the other end. The 3′ end is identified by an unbound (free) 3′ carbon atom (carbons atoms are numbered 1′thru 5′ in the sugar ring). The 5′ end is identified by an unbound 5′ carbon atom. When a double stranded nucleotide is formed, the 3′ region of one strand binds to the 5′ region of a complimentary strand (they are arranged in an antiparallel fashion). Figure 2.3 shows a 4-base sequence that could represent a very small segment of your DNA. Try

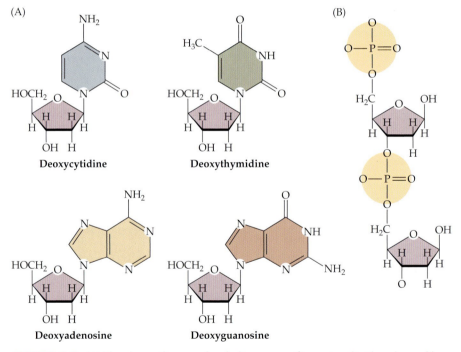

FIGURE 2.2 (A) The deoxyribose molecule has a specific region that bonds readily to any of the four DNA bases (adenine, cytosine, guanine, thymine) through a dehydration synthesis process. Similarly, each of the four bases has a specific region that bonds to deoxyribose. (B) The DNA backbone is created by the bonding of phosphate groups to deoxyribose molecules in a continuous chain. This structure is a common feature of DNA in all individuals and species.

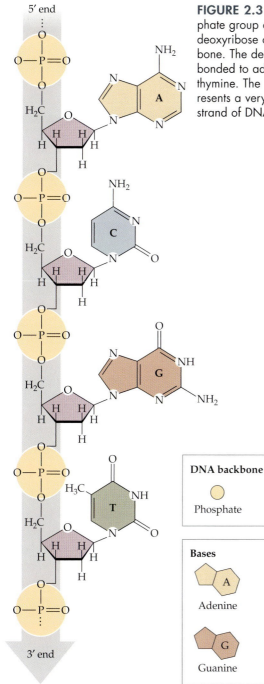

5′ end

3′ end

FIGURE 2.3 In a DNA molecule, the phosphate group of one nucleotide is bound to the deoxyribose of the next to form the DNA backbone. The deoxyribose molecules, in turn, are bonded to adenine, cytosine, guanine, or thymine. The sequence of bases shown here represents a very small segment of a much larger strand of DNA.

DNA backbone

Phosphate   Sugar

Bases

A   C

Adenine   Cytosine

G   T

Guanine   Thymine

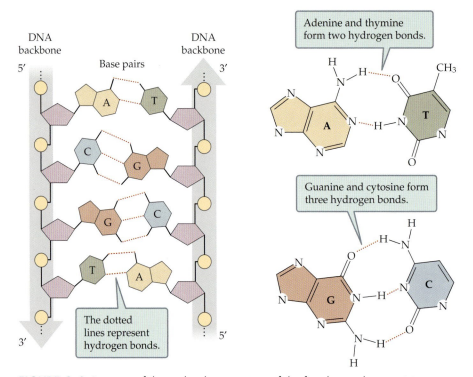

FIGURE 2.4 Because of the molecular structures of the four bases, base pairings are absolutely predictable: adenine always binds with thymine, and guanine always bonds with cytosine.

to imagine expanding this sequence from 4 bases to 40 bases, 40,000 bases, and then 40 million and beyond! In fact, a single strand of human DNA, known as a **chromosome**, contains more than 40 million bases.

A distinct feature of DNA is the variation in the sequence of bases along the length of each strand. Strands of DNA appear to contain millions of randomly assorted adenine, cytosine, guanine, and thymine molecules. Far from being random, however, the sequence of those bases is the essence of the DNA code used for production of every protein, cell, and structure of the organism. This code will be discussed in greater detail in the final section of this chapter.

If the process of assembling a DNA molecule has made sense to this point, the final step should be relatively easy to understand. DNA is not a single chain of nucleotides, but a double-stranded molecule. In essence, DNA consists of two separate but bonded strands, each with an identical sugar–phosphate backbone, but with their bases held together by hydrogen bonds. The base sequences of

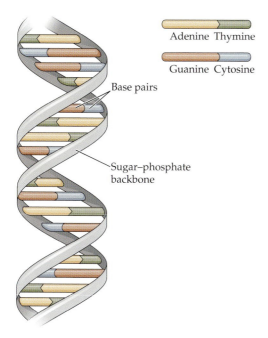

Adenine Thymine

Guanine Cytosine

Base pairs

Sugar–phosphate
backbone

FIGURE 2.5 If the sugar–phosphate backbone of DNA was formed with these molecules in a straight line, DNA would look like a conventional ladder (with the bases forming the "rungs"). Instead, the slightly angled nature of the sugar–phosphate bond results in a curving backbone that more closely resembles a spiraling staircase. The term typically used for this configuration in a two-stranded DNA molecule is the double helix.

the two strands are predictably arranged because they bond in a complementary fashion. The term **complementary base pairing** refers to the fact that each of the four bases can bind to only one other base. Specifically, adenine can bond only to thymine, and vice versa (A to T), and guanine can bond only to cytosine, and vice versa (G to C) (Figure 2.4).

The predictable pairing of bases occurs because of a molecular attraction between specific hydrogen bonding configurations shared by complementary base pairs. Guanine and cytosine each have three hydrogen bonding sites, whereas adenine and thymine each have two hydrogen bonding sites (see Figure 2.4). Because of the clear difference between these configurations, the complementary sequences are absolutely predictable.

At this point, the general structure of the DNA molecule should be fairly clear. It is held together by covalent bonds between phosphates and sugars, covalent bonds between sugars and bases, and finally, hydrogen bonds between complementary base pairs. One feature that is left to explain is the familiar spiral shape known as the double helix (Figure 2.5). The reason for this shape is fairly simple as well. The sugar–phosphate bond is not straight, but rather is formed with an angle. Because of this, the DNA backbone forms a continuous curve spiraling around itself. When two complementary strands are bonded together, the result is the famed double helix (Box 2.1).

BOX 2.1   Discovery of the Double Helix: Credit where Credit Is Due?

The double-helical structure of DNA may be simple, but the process by which that structure was discovered was not. Most people know that a Nobel Prize was awarded for its discovery. Some people even know that the award was given in 1962 to three researchers: James Watson, Francis Crick, and Maurice Wilkins. But few are familiar with Rosalind Franklin and her association with the discovery of the structure of DNA.

The story began in 1950, when Franklin, a young Ph.D. in physical chemistry, was doing pioneering work in X-ray crystallographic methods at King's College, London. Using X-ray beams to analyze DNA, Franklin found two forms of the molecule: an A form (dry and easy to photograph) and a B form (wet and difficult to photograph). Knowing that phosphates readily attract water, and that DNA was easily hydrated and dehydrated, she hypothesized that the DNA molecule had a double-helical shape, with a phosphate backbone on the outside and bases inside.

James Watson was in attendance when Franklin presented her findings and theories at a seminar in November 1951. Using information from that seminar and other sources, Watson worked with his colleague Francis Crick to develop a model of the DNA molecule. This original model was a triple helix with bases on the outside. In the spring of 1952, Franklin photographed the B form of DNA, clearly showing the double-helical structure she had hypothesized. However, rather than publishing her findings immediately, she worked for months on laborious calculations to determine whether the A form was similarly helical.

Early in 1953, Watson discussed his theories with Franklin's colleague Maurice Wilkins. During their discussions, Wilkins showed Watson one of Franklin's photographs without her knowledge or consent. Watson consulted with Crick about the photograph, and Crick, after reviewing Franklin's data, was able to hypothesize the double-helical structure.

Rosalind Franklin

In March 1953, Watson and Crick published their now famous paper describing DNA as a double-helical molecule with specific base pairings (Watson and Crick, 1953). The paper was the cornerstone of their Nobel Prize–winning research. Sadly, the only credit given to Franklin in the paper was the following statement: "We have also been stimulated by a knowledge of the general nature of the unpublished results and ideas of Dr. M. H. F. Wilkins, Dr. R. E. Franklin, and their co-workers at King's College London."

In 1958, at the age of 37, Rosalind Franklin died of ovarian cancer. In 1962, Watson, Crick, and Wilkins received the Nobel Prize in Physiology or Medicine. In their Nobel lectures, they did not include Franklin among their 98 references. Her name was listed in acknowledgements only by Wilkins.

This is a very brief account of the events surrounding the discovery of the structure of DNA. Much more information can be found in biographies of Watson, Crick, and Franklin, though many tend to be rather subjective. This story is important in part because it points out the necessity of cooperative research to ensure appropriate credit where credit is due. Some large research projects now actually list group names rather than individual names (i.e., The Human Genome Project, The Huntington's Disease Collaborative Research Group) in an effort to reduce quests for personal gain and to encourage data sharing and open, selfless collaboration. (Suggested reading: Sayre, 1975; Klug, 1968.)

Function

As mentioned previously, it is often said that DNA is the "blueprint for life." Such clichés usually have some factual basis, and this one is no exception. The base sequences of DNA specify the sequences of amino acids that make up proteins. Proteins, in turn, are required for the structure, function, and regulation of the body's cells, tissues, and organs. Furthermore, each type of protein has a unique function. **Enzymes**, for example, are specialized proteins that act as biological catalysts to regulate the speed of chemical reactions involved in metabolism. **Receptors** are membrane-embedded proteins capable of altering cellular activity when they interact with chemicals such as neurotransmitters, neuromodulators, hormones, or psychoactive drugs. Other examples of specialized proteins include antibodies and hormones. Thus, DNA possesses almost limitless potential to influence the physical development and functioning of an organism.

Most DNA resides in the nucleus of a cell, although there are some exceptions (Box 2.2). In addition to containing the cell's genetic information, the DNA

BOX 2.2   Mitochondrial DNA

Most people with a basic understanding of biology know that mitochondria are organelles responsible for converting sugars into the energy required for normal cell function. What sets mitochondria apart from other organelles is that they are self-replicating. Cells do not produce their own mitochondria; instead, new mitochondria arise within a cell through the division of preexisting mitochondria. Mitochondrial self-replication utilizes a sequence of DNA that is separate and distinct from the nuclear DNA we have described in this chapter.

Why do mitochondria have their own DNA? The predominant theory is that these organelles were once independent prokaryotes (cells without nuclei) that provided a useful function when they were taken up by eukaryotic cells (cells with a distinct nucleus containing nuclear DNA). That useful function, presumably, would have been their ability to more efficiently convert food into energy that could be utilized by the host cell. Through evolution, mitochondria were incorporated into eukaryotic cells as essential organelles. At the same time, they retained their ability to replicate their DNA and mitochondrial division does not necessarily coincide with the replication process of their host cell.

How much mitochondrial DNA is there? Human mitochondrial DNA consists of 16,569 base pairs and only 13 genes—many fewer than the approximately 3 billion base pairs and 20,000–25,000 genes in human nuclear DNA. The mitochondrial genome, however, contains little junk DNA; instead, nearly every part of the DNA molecule forms a part of the 13 genes. Despite its small size, the potential effects of mitochondrial genome on cell development, cell function, and ultimately on behavior should not be underestimated. Unfortunately, because researchers are currently in the early stages of understanding the contributions of genetics to behavior, mitochondrial DNA has received limited attention. It is likely that researchers will have to consider the influences of this separate realm of DNA as they begin to assess the genetic factors that influence behavior.

can be copied to pass that information on to the cell's descendants; in other words, it is **self-replicating**. When a cell divides, the hydrogen bonds between base pairs are broken, and the DNA molecule unwinds into two separate strands in a sequential pattern that resembles the opening of a zipper. The two strands then become templates for the synthesis of new, complementary strands. The end result is two daughter molecules, each an exact replica of the parent molecule (Figure 2.6). Each daughter molecule consists of an original strand (the template strand from the parent molecule) and a newly synthesized strand of DNA. Replication of DNA is a central component of cell division, and thus of the growth and maturation of organisms.

Where do the nucleotides required for DNA replication come from? The vast majority of these nucleotides are synthesized by the cell from carbohydrates and amino acids and remain free in the cell until they are needed. Not surprisingly, this means that in animals, the building blocks of DNA are contained in the basic elements of the diet.

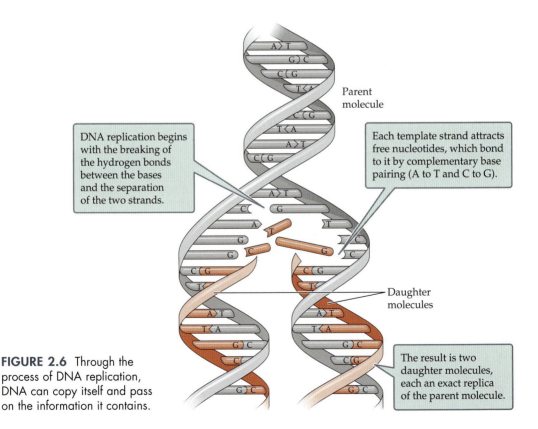

Parent molecule

DNA replication begins with the breaking of the hydrogen bonds between the bases and the separation of the two strands.

Each template strand attracts free nucleotides, which bond to it by complementary base pairing (A to T and C to G).

Daughter molecules

The result is two daughter molecules, each an exact replica of the parent molecule.

FIGURE 2.6 Through the process of DNA replication, DNA can copy itself and pass on the information it contains.

RNA

RNA is the abbreviation used for **ribonucleic acid**. RNA contains the same basic subunits as DNA: nucleotides made up of a five-carbon sugar, four bases, and a phosphate group. In the case of RNA, however, the sugar is ribose, not deoxyribose; in other words, no oxygen atoms have been lost from it. Like deoxyribose, ribose has specific regions that bond to phosphate groups and bases. Three of the bases it bonds to are the same as those found in DNA: adenine, cytosine, and guanine. RNA, however, does not contain thymine, but instead utilizes another pyrimidine, **uracil** (Figure 2.7).

Structure

The process of molecular bonding in RNA is the same as in DNA: sugars bond with phosphates and bases to form nucleotides, and phosphate groups bond with sugars to form a long chain. The end result is a single-stranded molecule consisting of a sugar–phosphate backbone and a sequence of the bases adenine, cytosine, guanine, and uracil. Adenine, cytosine, and guanine bind to ribose molecules just as they bond to deoxyribose molecules (see Figure 2.4). Uracil bonds to ribose molecules as shown in Figure 2.7.

Function

If DNA represents a blueprint for life, RNA may be thought of as the construction workers that utilize that blueprint for building structures. There are several forms of RNA that carry out different construction tasks. **Messenger RNA (mRNA)** is assembled within the cell nucleus, using DNA as a template, and then transported out of the nucleus into the cell cytoplasm. This process, called **transcription**, begins when a specific segment of double-stranded DNA within the nucleus unwinds, exposing its bases. **RNA polymerase**, an enzyme that catalyzes the synthesis of RNA, then transcribes the DNA base sequence into a single strand of RNA that forms

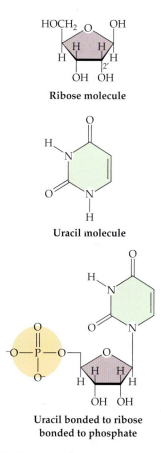

Ribose molecule

Uracil molecule

Uracil bonded to ribose
bonded to phosphate

FIGURE 2.7 Like DNA, RNA is made up of nucleotides consisting of sugars bonded to bases and phosphate groups. However, the sugar used in RNA is ribose, rather than deoxyribose, and RNA contains the pyrimidine uracil, rather than thymine.

TABLE 2.1	DNA–DNA and DNA–RNA Base Pairings	
DNA–DNA		**DNA–RNA**
adenine–thymine		adenine–uracil
thymine–adenine		thymine–adenine
cytosine–guanine		cytosine–guanine
guanine–cytosine		guanine–cytosine

the basis for mRNA. In this transcription process DNA guanine bonds to RNA cytosine, DNA cytosine bonds to RNA guanine, DNA thymine bonds to RNA adenine, and DNA adenine bonds to RNA uracil. While the first three base pairings will be familiar from the previous description of the DNA double helix, the fourth pairing is unique because uracil is not shared by DNA and RNA. In brief, just remember to substitute U for T in the RNA when describing base pairings that occur during transcription (Table 2.1).

Although DNA provides the template for RNA synthesis, it should be pointed out that only a fraction of the DNA in the nucleus of a cell is actually transcribed. Furthermore, only a small portion of the newly synthesized RNA becomes mRNA that ultimately leaves the nucleus. To understand why, a brief explanation of RNA splicing is needed.

The transcription process begins with a segment of DNA containing base sequences that code for a protein. These segments containing useful codes are called **exons** (Figure 2.8). In many cases, however (and for reasons still not com-

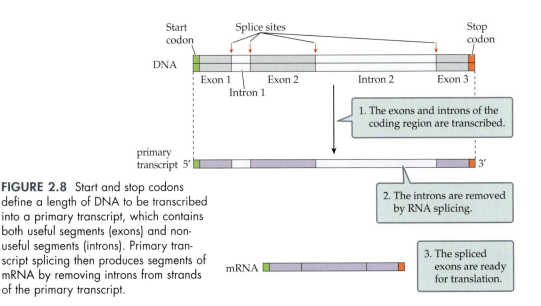

FIGURE 2.8 Start and stop codons define a length of DNA to be transcribed into a primary transcript, which contains both useful segments (exons) and non-useful segments (introns). Primary transcript splicing then produces segments of mRNA by removing introns from strands of the primary transcript.

pletely understood), exons are interrupted by lengthy stretches of noncoding DNA. Such DNA segments containing unusable base sequences are called **introns**. In the cell nucleus, a DNA segment that includes both exons and introns is first transcribed to create a complementary RNA known as a **primary transcript**. The process of **RNA splicing** occurs in a second step, in which introns are removed from nRNA. The final product is a strand of mRNA containing an edited sequence of exons. That mRNA is transported out of the nucleus and into the cytoplasm.

Once out of the nucleus, the mRNA attaches to **ribosomes**, the organelles where protein synthesis takes place. Another type of RNA molecule, **transfer RNA (tRNA)**, "reads" the mRNA base sequence (the message) and brings the amino acids it specifies to the ribosomes, where they are assembled to form a protein (Figure 2.9). This process, called **translation**, is the essence of cell development, literally constructing cells and altering their function to conform to the specifications of the DNA blueprint.

Later in this chapter, a more detailed description of the DNA and RNA code that specifies amino acids will be provided. At this point, it is sufficient to understand that this code is used to construct not only the basic shared features of all organisms, but also the features that make each individual unique. This simple code, consisting of only four bases, dictates all neural development, influencing factors as complex as human moods, emotions, and other behaviors. Furthermore, in understanding how this code may be passed from parent to offspring, researchers have found a basis for making predictions about their potential influence on future generations. Thus, to understand DNA is to understand the most elementary mechanism underlying behavior genetics.

Although DNA does indeed direct protein production, it is not the sole factor determining cell structure and function. As mentioned in the previous chapter, environmental factors may interfere with DNA instructions, particularly during the developmental process. Some factors that commonly affect human neural development include exposure to chemicals or drugs and nutritional alterations. To use the blueprint–construction worker analogy one more time, imagine having a fine blueprint and very talented workers, but that those workers show up for work drunk every day. How would this affect the final product? **Fetal alcohol syndrome (FAS)** is characterized by physical and behavioral abnormalities caused by exposure of the fetus to alcohol during critical periods of development. Features of FAS may include generally retarded growth, small head circumference, mental retardation, and impulsive behavior. Thus, even in the absence of defects in the genetic code, environmental influences can result in structural and functional deficits.

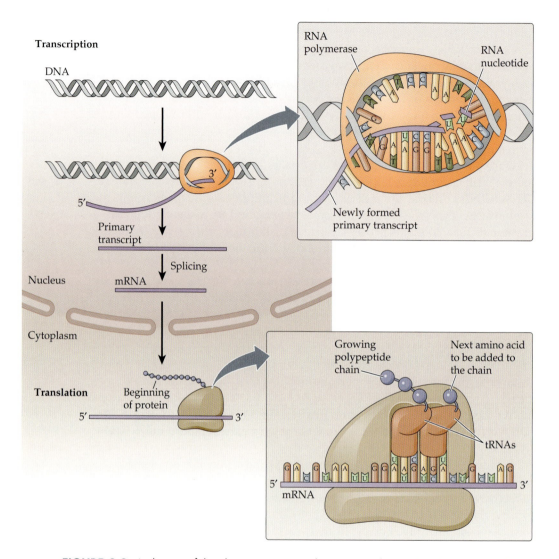

Transcription

DNA

3′

5′

Primary transcript

Splicing

Nucleus

mRNA

Cytoplasm

Translation

Beginning of protein

5′   3′

RNA polymerase

RNA nucleotide

Newly formed primary transcript

Growing polypeptide chain

Next amino acid to be added to the chain

tRNAs

5′   3′

mRNA

FIGURE 2.9 At the top of this diagram, DNA is shown unraveling within the nucleus, allowing complementary nucleotides to bind to it and form a sequence of primary transcript. The binding of nucleotides to form a primary transcript is catalyzed by RNA polymerase that recognizes a base sequence that initiates the process. Once formed, this single strand of primary transcript is then spliced to produce mRNA, which leaves the nucleus, carrying instructions for protein assembly into the cytoplasm. As mRNA attaches to ribosomes, tRNA binds to complementary base pairs on the mRNA. This process, called translation, strings together amino acids, specified by mRNA base sequences, into polypeptide chains that increase in size and complexity to form functional proteins. Between 1,000 and 2,000 bases must be bound by a ribosome to produce a typical protein molecule.

Chromosomes

Chromosomes are structures made up of long strands of intertwined DNA molecules. To gain some idea of how complex a chromosome might be, consider that even one of the smallest human chromosomes (chromosome 21) contains over 33 million base pairs (Figure 2.10).

A chromosome consists of two "arms" (Figure 2.11). These arms originate from the **centromere**, a region of DNA that does not contain genetic information, and extend to the **telomeres**, which are the end regions of the chromosome. Because the centromere does not contain genetic information, it has no direct role in the structural development or function of the organism; instead, it plays an important role in cell division. (Do not confuse the term "centromere" with the center of the chromosome. In more cases than not, the centromere is not at the midpoint of the chromosome.)

Chromosome

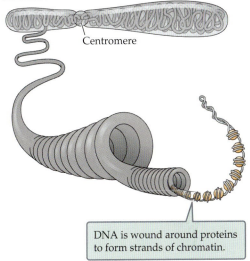

Centromere

DNA is wound around proteins to form strands of chromatin.

FIGURE 2.10 How are millions of base pairs packed into such a small structure? DNA is wound around proteins that are coiled into long strands. These strands are then tightly woven into a chromosome.

A major task in the study of chromosomes is mapping out specific regions associated with functional DNA codes. To do this, researchers must utilize reliable landmarks on each chromosome. The centromere provides one obvious landmark from which each chromosome may be visually subdivided. Extending from the centromere, each chromosome has a short arm and a long arm. The short arm is denoted **p** for "petite." The long arm is denoted **q**. Most historical sources suggest that q does not stand for anything in particular, and that researchers chose it only because it was the letter following p (not very creative, but easy to remember).

Diagrams and photographs of chromosomes typically show a distinct pattern of horizontal striping called **cytogenetic bands**. These bands are an artifact of staining techniques and may be seen as dark and light or florescent and nonfluorescent sequences (depending on the staining technique used). Cytogenetic bands do not appear to have any functional significance. They do, however, appear consistently in chromosomes: each of the 23 human chromosomes exhibits a distinctive and reliable banding pattern. Geneticists have numbered these bands for each chromosome, identifying them by arm and position relative to the centromere. Using this method, researchers subdivide chromosome

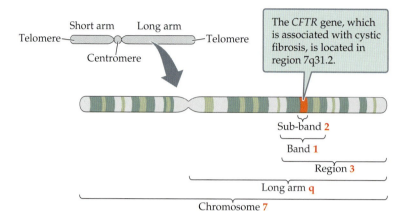

Chromosomal location: **7q31.2**

FIGURE 2.11 The bands and sub-bands of chromosomes create reliable landmarks that can be used for identifying gene locations, as shown in this diagram of chromosome 7. (After U.S. National Library of Medicine.)

arms into regions and then use bands, and in some cases **sub-bands** (less distinct bands within bands), to identify locations of interest, such as genes. For example, the gene associated with cystic fibrosis (**CFTR gene**) is found at 7q31.2: on the long arm of the chromosome 7, in region 3, band 1, and sub-band 2 (see Figure 2.11). The gene that is associated with Huntington disease (**HD gene**) is located at 4p16.3: the short arm of chromosome 4, region 1, band 6, sub-band 3. This system of notation based on reliable chromosome landmarks allows researchers to find regions of interest quickly.

The telomeres of the short and long arms of each chromosome are referred to as **ptel** and **qtel**, respectively. Referring back to Figure 2.11, it has been determined that at least one gene contributing to adult height is located on chromosome 7 (Hirschhorn et al., 2001), but its location has not been as precisely narrowed down as that of *CFTR* or *HD*. The location of this gene is described as region 7q31–7qtel: the area extending from region 3, band 1 to the tip of the long arm of the chromosome.

How many chromosomes are there in humans?

Humans normally have 46 chromosomes containing a total of about 3.08 billion base pairs. Later chapters in this text will address genetic abnormalities associ-

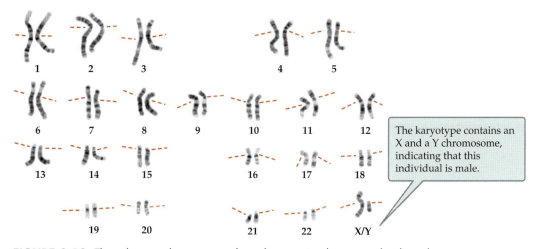

1 2 3 4 5

6 7 8 9 10 11 12

The karyotype contains an X and a Y chromosome, indicating that this individual is male.

13 14 15 16 17 18

19 20 21 22 X/Y

FIGURE 2.12 These human chromosomes have been stained, arranged in homologous pairs, and photographed to show an individual's karyotype. When chromosomes are visualized, they are usually shown in homologous pairs, with p on the top and q on the bottom. Note the obvious bands created in the staining process. Dashed lines indicate the locations of the centromeres for each homologous pair.

ated with additions to, or deletions from, this typical number. The full complement of chromosomes in a cell or an individual is referred to as its **karyotype**. The normal human karyotype of 46 chromosomes is generally represented as 23 homologous pairs (Figure 2.12). Twenty-two of these pairs of chromosomes, called **autosomes**, do not vary with the sex of the individual. The twenty-third pair, however, is made up of two **sex chromosomes**. In males, the sex chromosome pair consists of an X and a Y chromosome. In females, the sex chromosome pair consists of two X chromosomes.

To produce nongerm cells, chromosomes undergo **mitosis**, a replication and division process that results in daughter cells containing exact copies of all the chromosomes found in the parent cell (Figure 2.13). To produce **gametes** (sperm and egg cells), chromosomes undergo **meiosis**, a replication and division process that results in daughter cells that each contain one-half of the total number of chromosomes found in the parent cell (one chromosome from each homologous pair). In humans, gametes contain a total of 23 chromosomes (one from each of the 23 pairs). Because gametes have only half the complement of chromosomes, fusion of a sperm cell and an egg cell at fertilization results in a full complement of chromosomes in the offspring. Furthermore, this process ensures that half of the genetic material in the offspring is contributed by the

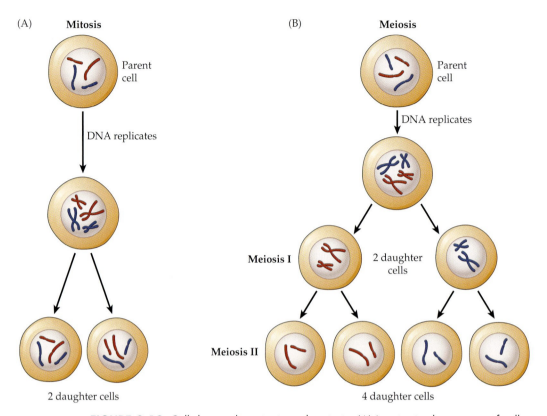

(A) **Mitosis**

Parent cell

DNA replicates

2 daughter cells

(B) **Meiosis**

Parent cell

DNA replicates

Meiosis I

2 daughter cells

Meiosis II

4 daughter cells

FIGURE 2.13 Cell division by mitosis and meiosis. (A) In mitosis, the process of cell division used by all nongerm cells, the chromosomes replicate, and the cell then divides into two genetically identical daughter cells. (B) The process of cell division used to create gametes also produces daughter cells in meiosis I. However, those two daughter cells then divide again without chromosome replication in meiosis II, resulting in four daughter cells containing one chromosome from each of the original homologous pairs in the parent cell. It should also be understood that segregation of chromosomes in Meiosis I is random, resulting in an equal probability for maternal and paternal (red and blue) to be integrated into each daughter cell. In the present example daughter cells received (by chance) the same colored chromosomes.

mother and half by the father. Because males have one X chromosome paired with one Y chromosome, half of the sperm cells produced by meiosis have a Y chromosome and the other half have an X chromosome. Because females have paired X chromosomes, all egg cells produced by meiosis have an X chromosome. This, of course, means that it is the genetic makeup of the sperm that ultimately determines the sex of the offspring.

How many chromosomes are there in nonhuman species?

Although we could probably argue that humans are among the most complex life forms in terms of behavior, humans are not necessarily the most complex in terms of genetic makeup. Species vary in their numbers of chromosomes and genes. As you might expect, the fruit fly has a very small number of chromosomes, with only 3 pairs of autosomes and 1 pair of sex chromosomes. Mice have a total of 20 pairs, and rats a total of 21 pairs. Considering the complexity of these animals, these numbers seem reasonable compared with our 23 pairs. But then consider that a horse has 32 pairs and a dog 39 pairs. Some might argue that, like humans, these mammals are highly evolved species that require many chromosomes for their complex development. It would be difficult, however, to use this logic to explain why a sweet potato has 45 pairs of chromosomes. In fact, the record for the number of chromosomes is held by a fern species with hundreds of pairs!

The point of comparing our own karyotype with those of species as diverse as fruit flies and ferns is twofold. First, it emphasizes that karyotypes may differ greatly among species. Second, it shows that the complexity of a species, and particularly the complexity of its behavior, is not directly correlated with its number of chromosomes.

Genes

Genes can be described in three ways. First, **genes** are the fundamental units of heredity that carry information from one generation to the next. Second, a gene can be described physically, as a segment of DNA in a particular location (referred to as a **locus**; plural **loci**) on a specific chromosome. Earlier in this chapter, for example, we referred to the *HD* gene responsible for Huntington disease as a segment of DNA located at 4p16.3 (on the short arm of chromosome 4). Finally, a gene can be described functionally, as the entity that is responsible for the synthesis of a specific product in the cell. The *HD* gene, for example, is known to be responsible for production of a protein called **huntingtin**.

A practical example may be helpful in thinking about this functional concept of a gene and its relationship to behavior. Some forms of clinical depression are believed to be associated with a deficiency in transmission of neuronal signals by the neurotransmitter serotonin. There is evidence to suggest that the deficiency is caused by dysfunctional serotonin receptors (Drevets et al., 2000). Because neurotransmitter receptors are proteins, their makeup is directly controlled by the cell's DNA blueprint. It is thus easy to imagine a blueprint error that could result in production of defective serotonin receptors. In this simple example, it is also easy to see how a defective gene could contribute directly to a behavioral abnormality. Finally, because genes are the fundamental unit of heredity, it is easy to see how some types of depression could be inherited (Cadoret, 1978; also see Hamet and Tremblay, 2005).

It is important to keep in mind the intimate association between genes and behavior. In the course of studying the myriad details of basic genetics, one can quickly lose sight of how this information applies to the study of human behavior. The following chapters describe a vast array of behaviors in terms of the influences of genes. Try to keep in mind while reading through those later chapters how directly genes can alter the structure and associated function of the cells that mediate those behaviors.

It should also be understood that the physical and functional descriptions of a gene become particularly important when researchers try to establish the source of genetically influenced behaviors. Pinpointing the locations of genes is a potentially powerful tool for identifying individuals who may be "at risk" for behavioral disorders. Conversely, determining which proteins are produced by a particular gene may offer clues needed to understand the biological basis of the behaviors associated with it.

Finally, it should be restated that behavior genetics is the study of the genetic basis for individual differences in behavior. As such, researchers are particularly interested in genes that vary in form and function. The term **allele** is used to denote an alternative form of a gene. Alleles underlie many observable differences between individuals. For example, there are three alleles of the gene that determines blood type (A, B, and O). The allele that produces Huntington disease is a form of the *HD* gene with an expanded nucleotide sequence, making it distinctly larger than a normally functioning *HD* gene.*

How many genes are there in humans?

Chapter 1 introduced the Human Genome Project, which was launched in the early 1990s. One of the goals of this project was the identification all of the genes in the human genome. It found that the DNA making up the normal complement of 46 chromosomes in humans contains between 20,000 and 25,000 genes.

Although this number may strike some readers as high, it is interesting to note that before the Human Genome Project began, most researchers in the field expected that humans would have at least 100,000 genes, and possibly many more. It became apparent early in the project, however, that the number of genes in humans would need to be revised downward. In fact, after the first decade

---

*In the wider literature about heritable traits, you will sometimes find instances of the term "gene" being used where "allele" is meant, as in "he inherited the gene for Huntington disease." Keep in mind that everyone has the HD gene, but only some have an allele of that gene that leads to disease.

of results from the project, most researchers cut their estimates in half. The project's initial reports estimated that the approximately 3 billion base pairs of the human genome contain approximately 30,000 genes (Venter et al., 2001). In 2004, the estimate was again revised downward to 20,000–25,000 genes (International Human Genome Sequencing Consortium, 2004). It is unlikely that the current estimates will be significantly revised, given the amount and quality of the data now available.

How many genes are there in nonhuman species?

The genomes of only a few animals have been mapped in enough detail to give a good estimate of the number of genes they contain. The fruit fly, for example, has a relatively large complement of about 13,500 genes embedded in its 4 pairs of chromosomes. Similarly, the simple roundworm *Caenorhabditis elegans* has about 19,000 genes in its 6 pairs of chromosomes. The mouse has 20 pairs of chromosomes, but its number of genes is estimated at only about 22,000.

Thus, like the number of chromosomes found in each species, the number of genes does not seem to be correlated with the complexity of an organism. What, then, is the factor that allows humans, with only slightly more genes than a roundworm, to develop into such complex organisms? The term **proteome** is used to refer to all the proteins that can be synthesized by a particular organism. It seems that the human genome allows for a much broader proteome than those of other species. This is because most human genes are believed to be capable of producing more than one protein product by means of **alternative splicing** (Modrek and Lee, 2002). In general terms, alternative splicing is the same as the splicing process described earlier for removing introns from the primary transcript. In this case, however, alternative splicing sites are used, allowing functionally variant mRNAs (and thus proteins) to be produced from the same DNA blueprint. Alternative splicing is thought to contribute to a greater potential for proteomic variation.

How much greater? Consider that the roundworm *C. elegans* has an estimated total proteome of 21,000 proteins, just slightly more than its total number of genes for that species (Takahashi et al., 2003). While the total number of proteins in the human proteome is still highly speculative, the range of estimates is typically between 500,000 and 1,500,000 (Hachey and Chaurand, 2004).

The field of **proteomics** is the study of proteins and their interactions within the cell. Proteomics is closely tied to the field of **genomics** (the study of genes) when assessing the influence of an organism's genes on protein production and ultimately on the structure and function of that organism. Clearly, humans represent one of the most complex organisms in terms of interactions between genes and protein production.

Gene Function

Try to conceptualize the entire physical being of an animal. What is it that constitutes skin, blood, bones, brain, and so forth? On a basic level, all these parts are made up of cells. But how are those cells made? A major component of cells is their proteins, which are large molecules composed of one or more chains of amino acids. To create a protein, these amino acids must be arranged in a certain order and folded into a particular shape.

What does a protein look like? Earlier in this chapter, the serotonin receptor was cited as an example of a protein that might be linked to a specific behavior. Serotonin neurotransmission is known to be closely associated with mood, and serotonin receptor proteins are an integral part of the neurotransmission process. Figure 2.14 is an example of what is commonly referred to as a **snakelike view** of a protein—in this case, the serotonin receptor. This diagram gives a two-dimensional view of the amino acid sequence and configuration that make up the protein.

Each amino acid incorporated into a protein, such as the serotonin receptor, is specified by a segment of DNA. The protein construction process is initiated by transcription, as described earlier, whereby a specific segment of double-stranded DNA unwinds, attracting complementary nucleotides that are then

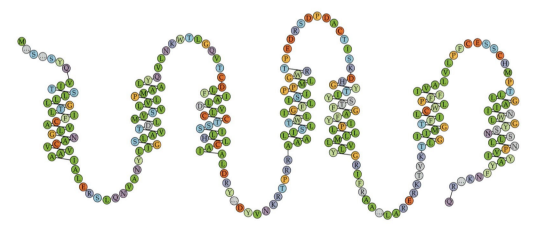

FIGURE 2.14 In this snakelike view of the serotonin receptor (1A subtype), each circle represents a particular amino acid. The sequence and the configuration of these amino acids combine to produce a functional protein. (After Campagne and Weinstein, 1999; Konvicka et al., 2000.)

bound together to create a strand of mRNA. The newly formed strand of mRNA moves out of the nucleus into the cell cytoplasm (see Figure 2.9). In the cytoplasm, with the help of tRNA and ribosomes, mRNA is translated into an amino acid sequence associated with a particular protein.

As we have seen, a gene is a segment of DNA that contains a specific sequence of base pairs used for construction of an amino acid chain that ultimately forms a particular protein. At this point, it should start to become clearer why DNA is referred to as the blueprint for life. But DNA also contains an abundance of additional information that does not appear to directly code for proteins. In other words, some of the DNA of a gene contains the vital blueprint information, but other DNA base pairs, both within and outside of genes, do not seem to influence protein production. Interestingly, over 95% of DNA does not appear to be transcribed or translated. This DNA is sometimes called **junk DNA**, but that term may be somewhat misleading. Because research into DNA and gene function is just beginning, it is possible that some of this junk DNA will be found to have regulatory functions.

What does the genetic blueprint look like?

Proteins are made up of 20 different amino acids. It would follow that if genes provide a blueprint for protein construction, there should be some reasonable order to the DNA sequences corresponding to these amino acids. This seems to be exactly the case. Each of the 20 amino acids is associated with a particular set of three-base sequences in the DNA code. Because binding of DNA base pairs to RNA base pairs is absolutely predictable (see Table 2.1), we also know the three-base sequence used by RNA to produce each of these amino acids. A three-base mRNA sequence associated with a particular amino acid is called a **codon**. Figure 2.15 lists the 61 mRNA codons associated with the 20 amino acids.

Protein assembly takes place at ribosomes with the help of tRNA (Figure 2.16). Each tRNA molecule has a three-base sequence that is complementary to an mRNA codon. A tRNA molecule binds to the amino acid specified by the codon it can bind, then travels to a ribosome and binds to that codon on an mRNA molecule. The amino acids lined up at the ribosome by tRNA molecules are then bound together into a polypeptide chain.

How are usable gene-length segments of RNA selected? The answer to this question lies in the codons that are used to initiate and terminate translation of mRNA. At least three codons used for amino acid production may also act as start codons to initiate translation. The sequence AUG is by far the most common **start codon**, being used to initiate translation in more than 90% of translated segments. CUG and UUG may also act as start codons, but together ac-

Review Questions and Exercises

■ Name the bases that make up DNA and list the complementary base pairings that are used in creating the DNA double helix. Name the bases that make up RNA and list the complementary base pairings between DNA and RNA.

■ Briefly describe the process of DNA self-replication.

■ The Huntington disease gene is identified as 4p16.3. Explain what the letter and each of the numbers in this abbreviated location description refer to.

■ Briefly describe the structure and function of a codon.

■ Define "proteome" and explain how the human proteome differs from the proteomes of other species.

■ Give an example of an environmental factor that could influence protein production and lead to behavioral dysfunction.

■ Explain what is meant by the term "genetic predisposition."

■ The Human Genome Project has revealed far fewer human genes than originally expected. Explain why this does not necessarily mean humans are less complex than previously believed.

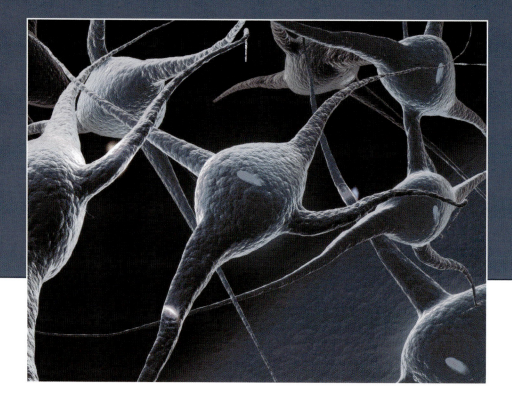

3

Neurogenetics and Neuropharmacology

The previous chapter introduced many basic concepts of genetics. Understanding that background information is the first step in understanding the complex relationship between genes and behavior. However, genes do not directly mediate behavior. Instead, they influence the development and function of structures that are responsible for producing behavior. The most obvious structure driving behavior is the brain, which is made up primarily of neurons and glial cells. Therefore, to understand the effects of genetics on behavior, one must understand the effects of genetics on neuronal function. This chapter outlines some basic concepts of neuronal function and neural communication. There is a particular emphasis here on those aspects of function and communication that may be directly altered through genetic changes. In addition, the chapter includes some background information on neuropharmacology, which is relevant for at least two reasons. First, neuropharmacology is the study of how drugs affect neural communication, ultimately influencing behavior. An understanding of how drugs alter neural communication processes provides important insights into how genetic factors might produce similar changes in neural communication and behavior. Second, neuropharmacology currently represents one of the fields of study that have been most effective in advancing therapeutic approaches to neural dysfunction. Research designed to enhance the understanding of genetic influences on both normal and abnormal neural communication could greatly enhance the potential of this already highly productive field of study.

FIGURE 3.2 Many ribosomes are located on the rough endoplasmic reticulum. The Golgi apparatus modifies proteins after their production at ribosomes and transports them to their final destination.

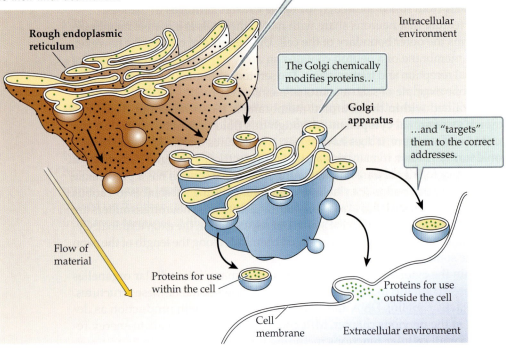

Proteins from the endoplasmic reticulum are transferred to the Golgi apparatus.

Intracellular environment

Rough endoplasmic reticulum

The Golgi chemically modifies proteins…

Golgi apparatus

…and "targets" them to the correct addresses.

Flow of material

Proteins for use within the cell

Proteins for use outside the cell

Cell membrane

Extracellular environment

Take a moment to think about all this information in terms of the influence of DNA on neuronal structure and function. Each neuron contains nuclear DNA that codes for production of the proteins—including the ribosomal proteins—that are ultimately responsible for the construction of all other proteins produced by the cell. In other words, DNA not only determines *what* proteins will be produced by a cell, but also all aspects of *how* those proteins will be produced by determining the structure and function of some organelles.

Neuronal function: Propagating a neuronal signal

To understand how a signal moves from one part of a neuron to another, it is necessary to look closely at some additional special properties of neurons. Most neurons maintain an electrical charge difference, called the **resting potential**, that is characterized by an imbalance of charged particles between the cytoplasm

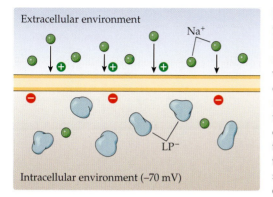

Extracellular environment

Na⁺

LP⁻

Intracellular environment (–70 mV)

FIGURE 3.3 A resting neuron maintains an electrical gradient and a concentration gradient between the cytoplasm and the extracellular environment. Large proteins (LP⁻) and nucleic acids with a strong negative charge are found inside the cell. Their negative charge contributes to the –70 mV resting potential of the neuron and is the basis for the electrical gradient that attracts positively charged sodium ions (Na⁺) into the cell. Because the concentration of sodium ions is higher outside than inside the resting neuron, those ions are also subject to a concentration gradient that, like the electrical gradient, moves them into the cell.

and the extracellular environment. More specifically, the intracellular environment maintains a negative charge of approximately –70 millivolts (mV) relative to the extracellular environment. This imbalance creates an **electrical gradient** that attracts positively charged ions into the cell. In addition, just as an electrical gradient attracts ions to a region with an opposite electrical charge, a **concentration gradient** attracts ions from an area of higher concentration to an area of lower concentration. Sodium (Na^+) is the most prevalent positively charged ion in the extracellular environment, where it is much more highly concentrated than in the intracellular environment. Therefore, movement of sodium from the extracellular environment into the neuron is facilitated by both the electrical gradient and the concentration gradient (Figure 3.3).

So what keeps the force of these two gradients from flooding the cell with sodium and eliminating the resting potential? The principal structure of the cell membrane is a **lipid bilayer**, made up of one layer of tightly packed lipid molecules lined up facing the extracellular environment, and a second layer lined up in the same tight formation facing the intracellular environment (Figure 3.4). The reason for this configuration is that the heads of lipid molecules are **hydrophilic** (attracted to water), while the tails of these molecules are **hydrophobic** (repelled by water). This particular arrangement results in a membrane that is impenetrable to all but the smallest particles (and a few substances that are highly soluble in lipids).

A neuronal membrane made up of only a lipid bilayer, through which few substances could move in or out, would result in a relatively inactive cell. The propagation of neuronal signals depends on ion exchange, so there must be some way for ions to pass through the membrane. They do so by moving through specialized proteins, called **ion channels**, embedded in the neuronal membrane, whose opening and closing is controlled by events in the cell (Figure 3.5).

FIGURE 3.4 Tightly packed lipid molecules form a nearly impenetrable cell membrane, allowing passage of only the smallest molecules and those that are soluble in lipids.

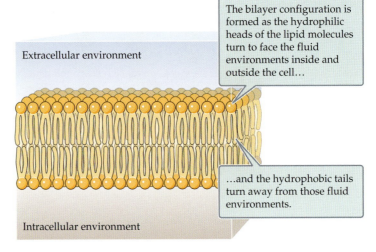

Extracellular environment

The bilayer configuration is formed as the hydrophilic heads of the lipid molecules turn to face the fluid environments inside and outside the cell…

…and the hydrophobic tails turn away from those fluid environments.

Intracellular environment

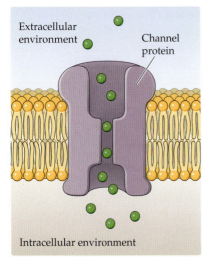

Extracellular environment

Channel protein

Intracellular environment

FIGURE 3.5 Ions that cannot pass through the lipid bilayer must pass through specialized channel proteins embedded in the neuronal membrane. These channel proteins, when opened, allow ions to flow freely into or out of the cell with the forces of electrical and concentration gradients.

To see how neuronal signals are propagated, imagine that a channel specialized for sodium transport opens in a neuronal membrane. The open ion channel allows an influx of positively charged sodium ions into the cell, caused by both the electrical and concentration gradients. This influx of sodium ions raises the electrical charge of the cytoplasm in the area around the channel from its normal resting potential of –70 mV. When the intracellular voltage is raised to a certain **threshold level** (typically –50 to –55 mV), it has the effect of opening **voltage-sensitive** ion channels located nearby in the cell membrane. These channels, in turn, allow further inflow of sodium ions, opening other adjacent voltage-sensitive ion channels and effectively creating a chain reaction that propagates the electrical charge increase along the cell membrane (**Figure 3.6**). The electrical signal created by voltage-sensitive ion channels is thus self-perpetuating, moving from channel to adjacent channel along the entire length of the cell membrane until it reaches the axon terminal.

When the influx of sodium ions reaches the axon terminal, the positive charge the ions carry stimulates

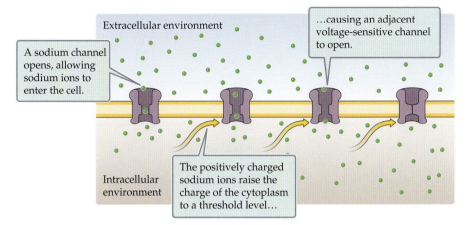

FIGURE 3.6 Voltage-sensitive channel proteins open when enough positively charged ions enter the cell to raise the charge of the cytoplasm to a threshold level. As ions enter one channel, they raise the charge of the intracellular environment around that channel to the threshold level, which in turn opens the adjacent channel. This arrangement allows a sequential and linear pattern of channel openings sometimes likened to a domino effect.

the opening of calcium (Ca^{2+}) channels. Like sodium ions, calcium ions flow readily into the cell with the force of both the electrical and concentration gradients. The function of calcium in the axon terminal, however, differs from that of sodium. Calcium inflow stimulates the movement of stored chemicals, packaged in vesicles, to the terminal membrane. Those vesicles may contain either of two types of neurochemicals: **neurotransmitters**, which are chemicals that directly alter the function of adjacent cells, or **neuromodulators**, which are chemicals that enhance or inhibit the effects of neurotransmitters (that is, they indirectly alter the function of adjacent cells). Calcium inflow moves the vesicles to a point where their membranes can fuse with the cell membrane, creating an opening into the extracellular environment. Through this opening, the vesicles release their contents from the axon terminal (Figure 3.7).

So what becomes of the neurochemicals released from the axon terminal? If the cell receiving these signals is a neuron, it will simply increase or decrease its activity in response to the signals. If the cell receiving these signals is part of a muscle, gland, or organ, the signal will evoke some functional response (i.e., muscle contraction or relaxation, hormone release, alteration of organ function).

At this point, the propagation of a neuronal signal has been described, but how that signal is initiated has not been explained. Recall that the neuron is a cell specialized for communication. The three basic steps in neural communication are **reception** (receiving a signal), **transduction** (converting a chemical

FIGURE 3.7 Whereas the sequence of voltage-sensitive sodium channel openings is used to transmit a signal along the length of the axon, it is the opening of calcium (Ca^{2+}) channels at the axon terminal that moves chemical-filled vesicles to the membrane, where they release their contents into the extracellular fluid, ultimately passing the signal to the next cell.

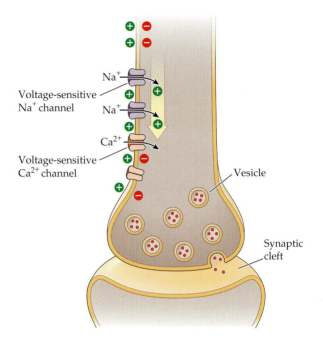

signal into an electrical signal, and vice versa), and **transmission** (moving the signal along the length of the neuron, usually via the axon, to another cell). Transmission was described above as the movement of electrical signals along the length of the cell membrane from the site of signal reception to the site where the neuron will pass the signal to an adjacent cell (typically from axon terminal to dendrite). To accomplish reception and transduction, the neuron uses specialized proteins embedded in its membrane. **Neurotransmitter receptors** are proteins whose activation by the binding of chemicals in the extracellular environment results in a functional change within the cell. Two basic types of receptors are involved in neuronal signaling, each of which will be described briefly in the next section.

Neuronal function: Initiating a neuronal signal

As their name implies, **ionotropic receptors** are physically and functionally attached to ion channels. These proteins bind neurotransmitters that are released from neighboring neurons. That binding induces the opening or closing of their associated ion channels and modifies the passage of ions through the neuronal membrane. This transduction of a chemical signal into a change in the electrical activity of a neuron is the essence of neuronal signal processing.

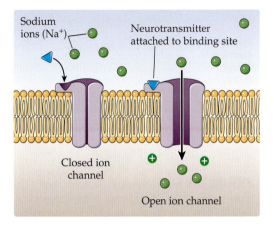

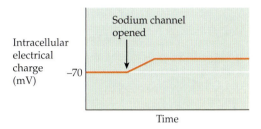

FIGURE 3.8 An ionotropic receptor that opens a sodium (Na⁺) channel in a neuronal cell membrane allows sodium ions to flow into the cell, resulting in depolarization associated with neuronal activation.

Like the ion channels described in the previous section, some ionotropic receptors allow positively charged ions to enter the cell. The resulting change in the electrical charge of the cytoplasm, from the negative resting potential to the more positive charge needed for initiating signal transmission, is known as **depolarization**. When a critical level of ionotropic receptor activity is reached, the influx of positively charged ions begins to open adjacent voltage-sensitive ion channels without the need for additional neurochemical stimulation (**Figure 3.8**).

Just as some ionotropic receptors initiate signal transmission by opening channels to allow the influx of positively charged ions, other ionotropic receptors are linked to channels that allow the influx of negatively charged ions (or the efflux of positively charged ions). When neurotransmitters bind to these receptors and the associated channels are opened, the electrical charge of the cytoplasm becomes more negative. This hyperpolarization (the opposite of depolarization) makes it more difficult for adjacent voltage-sensitive ion channels to open and ultimately slows neuronal signal transmission, a phenomenon known as **inhibition** (Figure 3.9).

Sometimes a concept such as the functioning of ionotropic receptors is more easily understood when tied to a practical example. Consider a person who suffers from a generalized anxiety or panic disorder. Try to imagine the neurochemical basis for such disorders. In some cases, a person with an anxiety disorder might have a dysfunctional ionotropic receptor controlling the influx of positively charged ions such that the channel opens more readily than normal. The dysfunction could be in the receptor protein, the ion channel, or both (recall that these proteins are created from a DNA blueprint). Thus, normal levels of neurotransmitters might produce abnormally high levels of neuronal activity in these individuals, resulting in overactivation of the sympathetic nervous

cerned with drugs that affect the nervous system. In particular, this section will focus on drugs that alter behaviors. Ideally, such drugs are intended for therapeutic applications, but the concepts discussed here also apply to recreational drug use and drug abuse.

The synapse

A basic understanding of neuropharmacology must begin with a general understanding of the synapse, where the chemical activity of interest takes place. The **synapse** is the junction, or gap, between the axon terminal of a neuron transmitting a signal (the **presynaptic** cell) and a portion of the membrane of the cell receiving the signal (the **postsynaptic** cell). The postsynaptic membrane is most commonly located on a dendrite of the postsynaptic cell. However, some postsynaptic membranes are found on other regions of neurons (including the soma, axon, and axon terminal) and on muscle, gland, or organ cells. As previously described, chemical communication between cells is generally initiated by the axon terminal, where vesicles containing neurotransmitters fuse with the membrane of the presynaptic cell, releasing those chemicals into the synapse (see Figure 3.7). The neurotransmitters then cross the synapse, a distance of about 200 angstroms (1 millimeter = 10,000,000 angstroms), to reach the membrane of the postsynaptic cell, where they bind to receptor proteins. After that initial binding, neurotransmitter molecules then proceed through a series of unbinding and rebinding events, with each binding initiating a new signal in the postsynaptic receptor protein.

A wide array of mechanisms can alter cell-to-cell signal transmission and transduction of neurochemical signals into electrical signals at synapses. For example, as mentioned above, neurotransmitter molecules released into the synapse and left undisturbed will repeatedly bind and rebind to postsynaptic receptors until they passively diffuse from the synapse. In a typical nervous system, however, they are seldom left undisturbed. More frequently, the neurotransmitter signal is effectively terminated by enzymes in the synaptic region that rapidly metabolize the neurotransmitter. One example of such enzymatic alteration is the cleaving of **acetylcholine** (**ACh**), an active neurotransmitter, into two inactive substances, acetate and choline. This process is accomplished by the enzyme **acetylcholinesterase** (**AChE**), which is readily available in the synaptic cleft (Figure 3.11).

Rapid deactivation can also occur through another active process, called **neurotransmitter reuptake**, that is initiated by proteins embedded in the presynaptic membrane. In this case, free-floating neurotransmitter molecules in the synapse are bound by a presynaptic membrane protein and then systematically

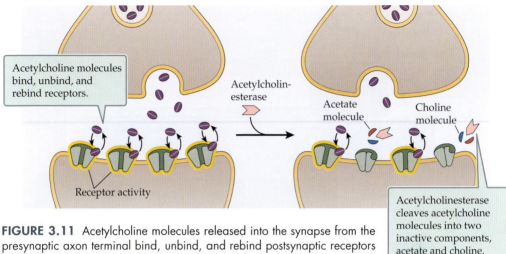

Acetylcholine molecules bind, unbind, and rebind receptors.

Acetylcholin-esterase

Acetate molecule

Choline molecule

Receptor activity

Acetylcholinesterase cleaves acetylcholine molecules into two inactive components, acetate and choline.

FIGURE 3.11 Acetylcholine molecules released into the synapse from the presynaptic axon terminal bind, unbind, and rebind postsynaptic receptors until they are deactivated by the enzyme acetylcholinesterase.

transported back into the axon terminal (**Figure 3.12**). This recycled neuro-transmitter can then be repackaged in terminal vesicles for future use. A simi-lar reuptake process may be used for inactive metabolites of neurotransmitters cleaved by enzymes in the synapse (e.g., acetate and choline). These substances are then transported back into the terminal, where they are reassembled into an active form of the neurotransmitter and repackaged in vesicles.

Autoreceptors are receptor proteins embedded in the presynaptic membrane that have the regulatory function of monitoring neurotransmitter levels in the synapse and reducing the release of additional neurotransmitter when critical levels are reached (see Figure 3.12). More simply stated, these receptors exert a negative feedback effect on further neurotransmitter release. Activation of these receptors has the effect of slowing neurotransmitter release, whereas blocking of these receptors can increase the amount of neurotransmitter avail-able in the synapse.

Understanding the mechanisms that regulate neurotransmitter activity in the synapse is a fundamental goal of neuropharmacologists. Identifying and designing drugs that alter these mechanisms is a primary application of this knowledge. The following sections will discuss some ways in which neu-ropharmacology has been used to alter behavior. While reading through the rest of this chapter, consider the many ways in which more information on the genetic basis of neuronal development and function could improve the field of neuropharmacology.

gression of cell death. Furthermore, the most likely target for neuroprotection is the synaptic region of the neuron, which appears highly vulnerable to degeneration (Coleman et al., 2004).

In addition to Alzheimer disease, other diseases that could benefit greatly from development of neuroprotectant therapies include Parkinson disease and Huntington disease. Although both Parkinson disease and Huntington disease have motor disturbances as their primary symptoms, the basis for these symptoms lies in chronic losses of cells in particular brain regions. Neuroprotectant therapies targeting these specific neurons could offer some relief for sufferers of these incurable diseases. However, such therapies will be difficult to develop by studying responses to pharmacological treatment of apparent deficits. Instead, most researchers agree that for neuroprotection to be effective, treatment will need to begin long before behavioral symptoms are apparent. Therefore, identifying genetic markers for neurodegenerative diseases will be a critical element of neuroprotection research, as these markers may be the best presymptomatic indication of disease development. Neuroprotectant therapies will be discussed in more detail in later chapters describing specific neurodegenerative diseases.

An introduction to pharmacogenomics

Neuropharmacological treatment of psychiatric disorders is arguably the most rapidly progressing area of both basic and applied research in the field of neuroscience today (though behavior genetics is rapidly closing the gap, particularly in basic research). Disorders such as major depression, chronic anxiety, obsessive–compulsive disorder, attention deficit hyperactivity disorder, and schizophrenia are all considered treatable using some form of drug therapy. This is not to suggest, however, that drug therapy represents an end point in research into these disorders. Even those individuals who are considered successfully treated with drugs must endure some initial period of drug adjustment, during which numerous types and doses of drugs may be tried. During this initial period, often lasting months, patients frequently experience recurrence of their symptoms or troublesome side effects. Even with an effective drug treatment regimen established, side effects may persist for as long as the drugs are taken. How long are these drugs taken? In many cases, patients must continue these pharmacological treatments for the duration of their lives.

With this in mind, drug research should be considered no more than an effective starting point in the treatment of psychiatric disorders. However, rapid advances in drug therapy for a vast array of disorders are likely to come from the union of neuropharmacological research with a greater understanding of genetics. In this approach, sometimes called **pharmacogenomics**, the genomes of individuals who suffer from a particular disorder are analyzed. The resulting information is further refined to determine the genetic makeup of those individuals who express a particular array of symptoms or, ideally, a positive response to a particular drug treatment.

Imagine for a moment the possibilities for implementing pharmacogenomics in general psychiatric practice. For example, there are currently dozens of drug treatments available for patients who report symptoms of chronic depression. Even after careful analysis of a patient's symptoms, however, psychiatrists must make an educated guess at which drug will be most effective. As mentioned above, drug treatment for depression may take weeks or months before significant benefits are realized. If, after this time, the psychiatrist and patient agree that the drug treatment is not effective, the process is repeated with a new drug. Considering the amount of research invested and the resulting information available, it would be unfair to use the term "shotgun approach" for this process of prescribing drugs. However, a fair description might be "educated trial and error."

Now consider the power of combining neuropharmacology with genetic research. Over time, genetic profiles could be collected from patients receiving treatment for depression. As the genetic database expanded, genetic markers shared by these patients would be identified. Then, in a final step, shared genetic markers would be identified among subsets of patients who responded positively to a particular drug or drug combination. With that information in hand, a genetic profile of a patient would be used to determine what drug the psychiatrist would initially prescribe, eliminating much of the initial guesswork.

This scenario is, of course, overly simplified, but it illustrates the point that if genes do underlie particular behavioral deficits, it follows that they underlie related neuronal dysfunctions. It should also be noted that skilled and observant psychiatrists may already be utilizing this technique to some degree, identifying the expression of aberrant gene combinations though clinical observations. For example, in an extensive review of the literature, Menza et al. (2003) found that psychiatrists frequently augment the core effects of prescribed antidepressants by prescribing drugs to treat additional symptoms not relieved by the initial drug or, in some cases, to treat side effects of the initial drug treatment. They concluded that treatment of these residual symptoms frequently improves the long-term outcome of drug therapy for depression. It is likely that pharmacogenomic research will eventually provide genetic profiles for many subclasses of depressed individuals (and individuals suffering from other psychiatric conditions). Such profiles would allow psychologists to more readily determine the optimal initial drug treatment protocol for a patient without the need to observe additional symptoms or side effects.

The application of pharmacogenomics to the treatment of several psychiatric disorders, most notably schizophrenia, has already begun. As the database of genetic markers for disease subtypes continues to grow, prescreening for such genetic markers may be a way of ensuring that the most effective and safest drugs are prescribed to patients. In this regard, pharmacogenomic research has the potential to vastly improve the application of neuropharmacological research, particularly for determining which psychoactive drugs should be initially prescribed.

Review Questions and Exercises

- List at least three organelles that are common to neurons and nonneuronal cells. Briefly describe the basic function of each of these organelles.

- Briefly describe the basis for both the electrical and the concentration gradients that influence ion flow into neurons.

- Briefly explain the difference between ionotropic and metabotropic receptors.

- Explain how genes that control neuronal development and function also, ultimately, influence behavior.

- List several ways in which either agonists or antagonists may exert their effects.

- Describe the function of presynaptic autoreceptors.

- Briefly explain how genetic markers could be used to improve drug therapy for psychiatric disorders.

- Give brief definitions of the fields of neuropharmacology and pharmacogenomics.

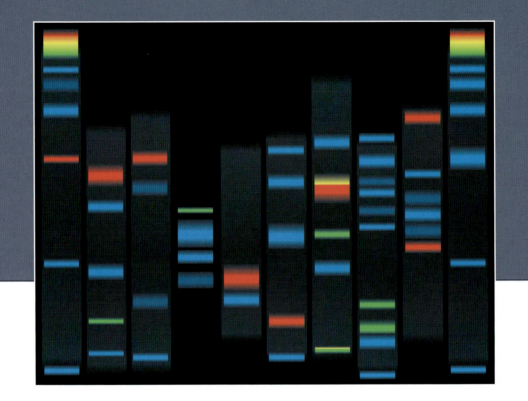

4

Methods of Genetic Discovery and the Human Genome Project

In the legends of King Arthur, the Holy Grail was a vessel said to have the power to heal all wounds and cure all illnesses. Because it was believed to possess these qualities, the grail subsequently became the object of extensive searching by King Arthur and his valiant Knights of the Round Table. Scientists sometime use the term "Holy Grail" to denote the pinnacle of accomplishment in any area of work or research. The term may be particularly appropriate, however, when discussing genetic research. Imagine the progress that could be made in medical and psychiatric treatment if researchers fully understood the human genome. Imagine the power of knowing which gene directs the production of each protein and how gene abnormalities alter protein synthesis and function. More than any other field of knowledge, genetics seems to hold a literal Holy Grail. So how far have scientists advanced in their search? How close are we to reaping the healing powers of genetic knowledge? A lot of work is still needed, but important results have been accumulating at a rapid pace, particularly over the past two decades. Perhaps the most intriguing recent advances in the study of genetics have been in the techniques used to identify specific features of DNA. These methods of discovery allow researchers to visualize genes and genetic abnormalities. Among the most impressive applications of these techniques has been the collaborative effort of researchers around the world to determine the complete sequence of the human genome. Before discussing the Human Genome

Project, however, it is worth taking some time to explore, at least on a rudimentary level, the most popular methods used in genetic discovery.

Visualizing DNA

DNA is small. Perhaps that is stating the obvious, but it is a reasonable starting point for describing the techniques required to work with it. If one wishes to study, in detail, something that is small, a method is needed for enlarging it. A magnifying glass, a microscope, or a telescope can be used to magnify the physical image of some objects, but there is no current technology available to do this for DNA. The process for enlarging a sample of DNA is somewhat different. Remember that DNA is a sequence of base pairs that readily replicates. With this understanding, researchers determined that the best way to "look at" a particular segment of DNA was to isolate that segment and then induce it to replicate rapidly and expansively. Although this approach does not enlarge the physical image of DNA, it does amplify its physical features, making them easier to detect and manipulate for analysis.

The polymerase chain reaction

The **polymerase chain reaction** (**PCR**) is a procedure used to amplify DNA samples by creating tens of millions of copies of a specific DNA segment. There are a number of reasons why researchers might be interested in a particular segment of DNA, but PCR is typically used when they wish to find similarities or differences between DNA samples. For example, if blood were found at a crime scene, or a man wished to gain paternity rights to a child, DNA samples from the blood at the crime scene and that of a suspect, or from the man and the child, would be amplified and assessed for similarities. On the other hand, if researchers wished to show that a disease had a genetic basis, they would hope to find differences between the DNA of affected and unaffected individuals. Keep in mind as you read through the description of PCR that the purpose of this process is simply to "view" a segment of DNA and compare it with other DNA segments, with the objective of establishing similarities and differences.

To begin the PCR process, a sample containing DNA is required. Such a sample could come from a variety of sources including blood, semen, saliva, or skin. Next, a specific segment of DNA from a particular chromosome is selected for study. That segment must lie between two **flanking regions**: segments of DNA for which the base sequences are known (Figure 4.1). A later section of this chapter will explain how scientists can identify base sequences that can be used as flanking regions for the DNA segments they wish to study. For now, just keep

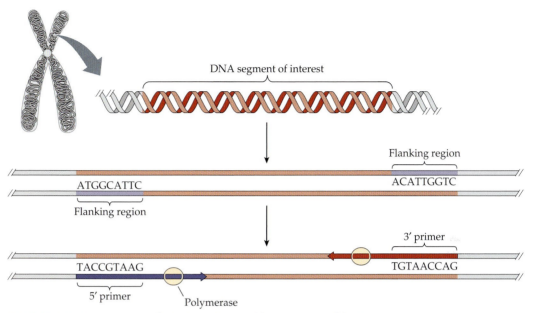

FIGURE 4.1 Any segment of DNA can be amplified with PCR if flanking regions can be identified. Flanking regions are segments of DNA whose base sequence is known. Primers that are complementary to these flanking regions can bind to them to initiate DNA replication from both directions through the region of interest.

in mind that researchers are able to identify such sequences throughout much of human DNA.

Once the segment to be studied has been identified, the DNA sample is incubated in a buffer solution that contains three key components:

1. **Primers**: short base sequences (typically less than 30 bases) that are complementary to the flanking regions
2. High concentrations of the four bases (A, T, C, G) required for assembling DNA
3. A **polymerase**: an enzyme that catalyzes the assembly of bases on a strand of DNA

Once a DNA sample has been mixed with the buffer solution, the remainder of the PCR process is relatively simple, requiring only three steps of heating and cooling (Figure 4.2):

- In the first step, called **denaturation**, the buffer is heated to 90°C–95°C for 30–60 seconds. Temperatures in this range break the hydrogen bonds

FIGURE 4.2 PCR is a three-step process, typically repeated approximately 30 times. This process, which can amplify a DNA sequence by approximately 1 billion times, typically takes less than 3 hours to complete. The simplicity and efficiency of this method have made it a primary contributing factor to the rapid advances in genetic research since its introduction in the 1990s.

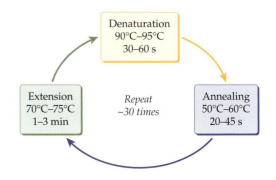

that create the DNA double helix by holding base pairs together. Denaturation thus produces individual strands of DNA.

- In the second step, called **annealing**, the buffer is cooled to 50°C–60°C for 20–45 seconds. At these temperatures, the primers bind (anneal) to the flanking regions of the denatured (separated) DNA strands. One primer, denoted 5′, initiates the assembly of new DNA inward from one flanking region of one of the denatured stands. The other primer, denoted 3′, initiates the assembly of new DNA inward from the opposite flanking region on the other denatured strand.

- In the third and final step, the buffer temperature is raised to 65°C–75°C for 1–3 minutes. This temperature range is ideal for the process of **extension**, during which the polymerase catalyzes assembly of complementary bases on each DNA strand, starting at the primers. The extension process moves in one direction from the 5′ primer and in the opposite direction from the 3′ primer (**Figure 4.3**).

Figure 4.3 may seem a bit overwhelming, but do not let the details of the process overshadow or confuse the basic concept of PCR. Look once again at Figure 4.2. The premise of PCR is not complicated. In its simplest form, it is a three-step process used to replicate a small segment of DNA.*

The final feature of PCR is repetition of the three steps. After allowing sufficient time for the third step, the buffer is heated again to the high temperatures required for denaturation, which also halts the process of extension. At this point, a new cycle of replication begins. In each subsequent cycle, the number of strands being replicated increases in exponential fashion, from 2 to 4 to 8 to 16 and so on. As **Figure 4.4** shows, over 1,000 copies of the DNA segment of interest are produced from a single strand of DNA in the first 10 cycles. In a typical PCR

---

*Max Animations (www.maxanim.com) has created a simple, yet educational, short animation depicting PCR. Viewing this animation at www.maxanim.com/genetics/PCR/PCR.htm may help interested readers to envision the basic processes of PCR.

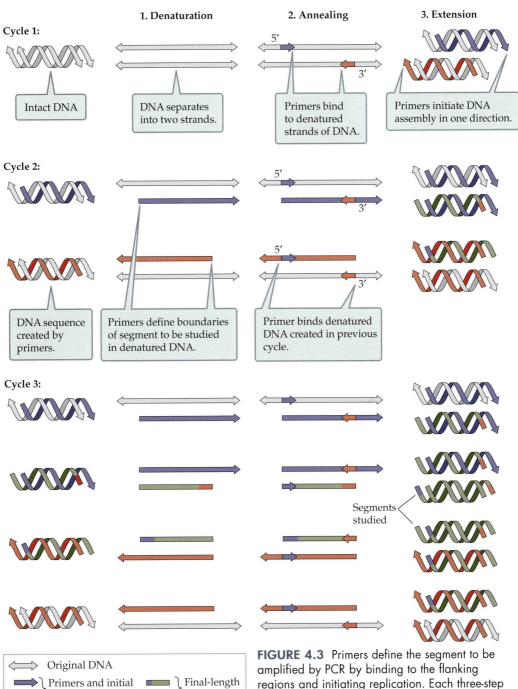

1. Denaturation

2. Annealing

3. Extension

Cycle 1:

Intact DNA

DNA separates into two strands.

Primers bind to denatured strands of DNA.

Primers initiate DNA assembly in one direction.

Cycle 2:

DNA sequence created by primers.

Primers define boundaries of segment to be studied in denatured DNA.

Primer binds denatured DNA created in previous cycle.

Cycle 3:

Segments studied

Original DNA

Primers and initial extension products

Final-length products

FIGURE 4.3 Primers define the segment to be amplified by PCR by binding to the flanking regions and initiating replication. Each three-step cycle doubles the number of copies.

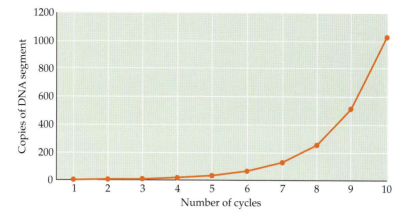

FIGURE 4.4 The amplification of a DNA segment by PCR is exponential. Just 10 cycles can generate more than 1,000 copies of the segment of interest. Over 30 cycles, approximately a billion copies of the segment are produced.

process, the cycle is repeated about 30 times. Each three-step cycle typically takes less than 5 minutes. Thus, the entire process takes only 2 to 3 hours to complete, yet typically yields over a billion copies of the DNA sequence of interest!

This section began with the statement that DNA is small. After reading though the description of PCR, you may be struck by the fact that, as small as it is, a sample of DNA can be readily amplified a billionfold in a relatively short time using this rather simple method. However, at least two questions remain unanswered at this point. First, how do researchers know what base sequences should be used as primers for PCR? In other words, if researchers wish to view a specific DNA sequence on a particular chromosome, how do they find short flanking sequences that occur reliably on both sides of that sequence? This information is rapidly becoming available though the Human Genome Project and related research. Scientists are currently working to identify all of the base pair sequences found in all of the chromosomes of a number of animal species, including humans. Although there is obviously some variation between individuals, many stretches of DNA are very similar or identical across individuals. The Human Genome Project has established a catalog of sorts, which will contain all of these shared DNA sequences for each human chromosome. The processes and problems associated with this enormous challenge will be covered in the second half of this chapter.

A second question left unanswered has less to do with the methodology of PCR than with its applications. What useful purpose do researchers have for a billion copies of a DNA sequence? It was noted earlier that PCR is used primarily to identify similarities or differences between DNA samples. The next section explains how.

Electrophoresis

The most common method used to visualize a sample of DNA that has been amplified using PCR is called **electrophoresis**. To begin the process, a thin block of gel is created, usually from **agarose** (a polymer typically extracted from seaweed). The agarose gel looks relatively solid, but in fact has microscopic pores, through which small substances (such as DNA fragments) can travel. Conveniently, DNA carries a negative electrical charge, which means that it is attracted to a positive electrical current. If you have followed the process to this point, the remainder of the explanation, which describes simple physics, is relatively straightforward. DNA samples to be compared are stained (marked with a highly visible dark or fluorescent dye). Small indentations, or "wells," aligned at one end of the agarose gel block are filled with solutions containing the labeled samples. A low-voltage electrical current is then run through the gel, with the positive pole of the current at the end of the gel block opposite the wells. The positive current slowly draws the DNA fragments from the wells though the pores of the gel, usually over a period of several minutes to several hours. The fragments create **bands** (not to be confused with cytogenetic bands, discussed in Chapter 2) at various points in the gel, depending on how far they have been drawn toward the positive pole (Figure 4.5).

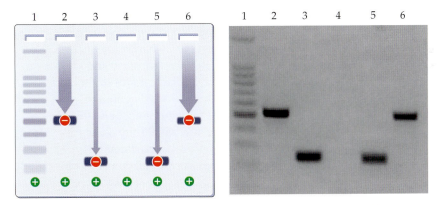

FIGURE 4.5 Electrophoresis is the most common method of visualizing the physical features of amplified samples of DNA. Wells (located at the top of the figures) at one end of an agarose gel block are filled with a solution containing the stained DNA samples. A positive electrical potential is applied the opposite end of the gel block (bottom of the figures). Negatively charged DNA is then drawn through small pores in the gel. Large fragments of DNA move relatively slowly through the pores (samples 2 and 6). Smaller samples travel through the gel more easily, moving a greater distance in the same amount of time (samples 3, 5). Sample 1 is a "ladder" used to create reference bands. Well 4 probably contained no sample, or contained a sample with stained fragments so small that they were drawn through the gel quickly and "ran off" the gel block, leaving no image on the gel.

Electrophoresis has two important features that make it useful. First, recall that each of the samples contains a billion or so identical fragments of DNA produced by PCR. Although any DNA fragment that is present in the sample will be stained and travel through the pores of the gel, when millions of identical fragments (the amplified signal) travel at the same rate they carry, by far, the greatest concentration of stain, making it easy to distinguish this sequence from any background (unamplified) DNA. The second feature is even more useful: large fragments of DNA have more difficulty moving through the pores, so they travel though the gel block more slowly than small fragments.

Think carefully about this simple concept while considering a practical example. Recall from Figure 4.1 that a segment of DNA can be defined by flanking regions. Imagine for a moment that you are performing electrophoresis to determine whether an individual has inherited the allele for Huntington disease. This allele is characterized by a triplet repeat expansion—that is, an excess number of repetitions of a three-base-pair sequence; in this case, CAG. For example, an unaffected individual generally has 35 or fewer CAG repeats in a known segment of DNA on the short arm of chromosome 4. In contrast, a person who inherits the Huntington disease allele has 37 or more—in some cases more than 100—CAG repeats (Bates and Lehrach, 1994). If you use primers for your PCR that flank the repeat region, then the more repeats, the bigger the amplified DNA segment. Because this triplet repeat expansion represents the basis for the disorder, it can be used to diagnose the presence of the disease. In addition, because the number of extra CAG repeats is typically correlated with the onset and severity of the disease, knowing the relative increase in the size of the DNA segment could offer some indication of how the disease will progress.

Figure 4.6 shows a hypothetical electrophoresis gel assessing DNA from a family in which the father showed behavioral effects of Huntington disease at the age of 44. Note that there are two bands on the gel block for each of the individuals tested. The reason for this is that PCR amplifies the same segment of DNA from both copies of chromosome 4 (recall that chromosomes are paired). The Huntington disease allele is carried on only one chromosome from the pair. The other chromosome carries a normal (unaffected) allele of the Huntington disease gene. Thus, in the case of the father, the band at the far end of the gel (farthest from the wells) indicates the smaller (unaffected) segment of DNA. The band in the middle of the gel indicates a larger strand of DNA that moved more slowly through the pores of the gel because of its size. Its larger size, of course, was the result of the CAG repeat expansion characteristic of the Huntington disease allele. The mother and son 1, both unaffected, produce two bands near the far end of the gel. In fact, son 1 appears to have a single band and the mother a single, somewhat larger band. This pattern would be expected if both chromosomes had identical or nearly identical segment sizes. The bands for both the

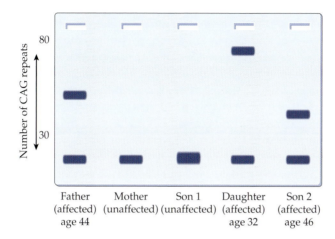

FIGURE 4.6 A hypothetical electrophoresis gel showing characteristics of DNA from the Huntington gene region of five members of a family in which the father carries the Huntington disease allele. Because PCR amplifies the sequence of interest from both copies of a chromosome, affected individuals (father, daughter, son 2) show two distinct bands. Bands farther from the wells indicate small DNA segments (which travel more easily, and thus move a greater distance, through the pores of the gel). Bands closer to the wells indicate large DNA segments whose increased size is attributed to the Huntington disease allele located on one chromosome. Note that the unaffected individuals appear to have a single band because the DNA fragments from both of their chromosomes (neither of which had CAG repeat expansions) traveled the same (son 1) or nearly the same (mother) distance. Finally, this gel shows that a greater number of CAG repeats (larger segment size) is associated with an earlier onset of the disease (as seen in the daughter).

mother and son 1 are in a region of the gel considered within the normal range of CAG repeats. Son 2 shows a CAG repeat expansion approximately the same size as his father's, and his age at the onset of the disease, as expected, was approximately the same as his father's. Finally, the daughter has a band close to the wells, indicating very little movement. This band suggests a CAG repeat expansion much greater than her father's or her brother's. As expected, the daughter's greater CAG repeat expansion was associated with an earlier age of onset of the disease—more than 10 years earlier than her father's.

DNA fingerprinting

Although the identification of disease-causing genetic aberrations is clearly a practical use of PCR and electrophoresis, these same procedures can also be useful for identifying regions of DNA that are not responsible for diseases, or even for normal behaviors. **Polymorphisms** are differences in any part of the DNA sequence between individuals. They may be, but are not necessarily, associated with differing phenotypes. One practical application of polymorphism analysis is the positive identification of individuals using blood or tissue samples that contain DNA. To do this, researchers utilize a **restriction enzyme**, an enzyme that cleaves a DNA strand at a particular base sequence. For example, *Eco*RI is a bacterial enzyme that cleaves DNA wherever the sequence GAATTC occurs. As mentioned in Chapter 3, much of our DNA is not directly associated with gene function. It is particularly within these lengthy nonfunctional segments of "junk DNA" that many polymorphisms occur. Consider that one individual's chromosome 20 may have far more GAATTC sequences (with no functional basis) than another individual's chromosome 20, or that they may have nearly the same number, but in different locations. Such patterns make GAATTC a useful polymorphism for individual identification.

Now tie together the concept of polymorphisms with the two methods that have already been described in this chapter: PCR and electrophoresis. Remember that PCR is used to amplify a particular segment of DNA so that its physical features can be visualized using electrophoresis. Restriction enzymes can be used to cleave a large segment of DNA into smaller segments of DNA that vary in size based on the characteristics of specific polymorphisms. Because the characteristics of polymorphisms vary widely among individuals, these smaller segments of DNA also vary in both size and number from one individual to the next. Thus, from any given DNA sample, polymorphisms can be analyzed using PCR and electrophoresis to create a distinctive DNA profile of an individual—a **DNA fingerprint**.

For a practical example, imagine that researchers wish to compare two samples of DNA to see whether they come from the same or different individuals. They perform PCR on both samples using primers that amplify a particular segment of DNA from a region on chromosome 20. Once they have produced a billion or so copies of this region from each sample, they use a restriction enzyme such as *Eco*RI to cleave each of those copies wherever the base sequence GAATTC occurs. If the amplified segments are polymorphic, the enzymes will cleave them into fragments of different sizes, which can be distinguished by using electrophoresis. This procedure could have four basic outcomes:

- First, if neither DNA sample contains the GAATTC sequence, two bands appear in the same location on the gel (recall that the primers used to generate the samples will have created fragments of approximately equal size) (Figure 4.7A). In this case, GAATTC is not a useful polymorphism for identification purposes, as the two profiles look very similar.

- The second possibility is that the DNA samples contain an equal number of GAATTC sequences (for simplicity's sake, assume that only one of these sequences appears in each sample). In this case, the DNA fragments, which were approximately the same size before exposure to the restriction enzyme, are each cut into two fragments. If the GAATTC sequence occurs in the same place in both DNA samples, two bands (representing the two fragments) appear for each sample on the gel, and they appear in similar locations (Figure 4.7B). Again, this is not a useful polymorphism for identification purposes.

- However, if the GAATTC sequence occurs in different locations within each sample, then the two bands for each sample appear in *different* locations. This third scenario represents a useful polymorphism for DNA fingerprinting.

- The fourth possible scenario is that the GAATTC sequence appears more frequently in one sample than in the other. If this is the case, the sample that is more frequently cleaved will produce a greater number of bands (Figure 4.7D). This DNA fingerprint is also useful for identification.

Note that in the useful scenarios of DNA fingerprinting, the numbers or locations of the bands appearing on the gel differ between individuals. It is easy to imagine an analysis of a single segment of DNA producing dozens of readily visualized variations that could be used to distinguish one individual from another. Adding a second segment of DNA to the analysis exponentially increases the number of possible differences between individuals. Researchers typically use five to ten different segments (sometimes more) to produce a DNA fingerprint, so the likelihood of any two individuals having identical fingerprints is very small. In fact, the probability of two identical DNA fingerprints occurring at random is frequently listed at less than 1 in 7 billion. With a world population of less than 7 billion, DNA fingerprinting is often considered conclusive proof of identity. Its use in law enforcement, however, has been subject to unforeseen problems and complications (Box 4.1).

Further applications of DNA analysis

The concepts of DNA fingerprinting are certainly interesting and have some application to the content of this text. However, other applications for visualizing DNA are more directly relevant to behavior genetics. Take the example of genetic diseases that are associated with a particular allele. Once a polymorphism has been identified as occurring reliably on a specific gene, PCR and electrophoresis become powerful tools for identifying the aberrant allele. Such polymorphisms have, of course, been identified for several prevalent single-gene

as possible to ensure optimal coordination of efforts among all participants in the project. For such a huge scientific undertaking, this kind of cooperation was essential. In fact, the project was expected to take 15 years, but cooperative and diligent work, combined with rapidly advancing technology, allowed the researchers to achieve their original goals two years ahead of schedule. Six goals were established at the outset of the project:

1. Identify all genes in human DNA
2. Determine the base sequences that make up all human DNA
3. Store all information in databases
4. Continually improve tools and methods for data analysis
5. Make data and technologies available to the private sector
6. Continually address ethical, legal, and social issues arising from the project

The first reports in 2001 came from two groups: Celera Genomics, a private company in Rockville, Maryland, and a second group of 18 institutions known as the International Human Genome Sequencing Consortium (Table 4.1). Although neither group claimed to have catalogued the complete sequence of the human genome, both reported similar and highly interesting findings. Among those findings was that the approximately 3 billion base pairs of the human genome contain only 20,000–25,000 genes. This number of genes is particularly interesting when compared with the numbers of genes found in much simpler organisms. Human genes, however, generally code for more than one protein product. As noted in Chapter 2, these findings suggest that human complexity is based in the proteome, rather than in the genes themselves.

Among the most intriguing implications of the downward revision of the number of human genes is that the amount of DNA making up our 25,000 genes represents only about 1% of our total DNA. Recall from Chapter 2 that the term "junk DNA" is sometimes used for DNA that does not contribute to genes. Does this finding imply that human DNA is approximately 99% "junk" and only 1% functional? It is clear that findings from the Human Genome Project will stimulate vast speculation about the origin and function not only of genes, but also of these apparently unused portions of DNA.

Data from the Human Genome Project suggest that the average human gene contains four exons (the coding regions within a gene), totaling 1,350 base pairs. The data also suggest that humans (and presumably most vertebrates) have genes that are not found in invertebrate species. Furthermore, a large amount of DNA appears in repetitive sequences. These repetitive sequences have presented a major obstacle to assembling the DNA sequences in proper order. One analogy for this problem is working on a jigsaw puzzle in which many pieces are the same color.

TABLE 4.1 **The Original 18 Institutions Contributing to the International Human Genome Sequencing Consortium**

THE SIXTEEN INSTITUTIONS FORMING THE ORIGINAL INTERNATIONAL HUMAN GENOME SEQUENCING CONSORTIUM:

1. *Baylor College of Medicine*, Houston, Texas, USA
2. *Beijing Human Genome Center*, Institute of Genetics, Chinese Academy of Sciences, Beijing, China
3. *Gesellschaft für Biotechnologische Forschung mbH*, Braunschweig, Germany
4. *Genoscope*, Evry, France
5. *Genome Therapeutics Corporation*, Waltham, Massachusetts, USA
6. *Institute for Molecular Biotechnology*, Jena, Germany
7. *Joint Genome Institute*, U.S. Department of Energy, Walnut Creek, California, USA
8. *Keio University*, Tokyo, Japan
9. *Max Planck Institute for Molecular Genetics*, Berlin, Germany
10. *RIKEN Genomic Sciences Center*, Saitama, Japan
11. *The Sanger Institute*, Hinxton, UK
12. *Stanford DNA Sequencing and Technology Development Center*, Palo Alto, California, USA
13. *University of Washington Genome Center*, Seattle, Washington, USA
14. *University of Washington Multimegabase Sequencing Center*, Seattle, Washington, USA
15. *Whitehead Institute for Biomedical Research*, Massachusetts Institute of Technology, Cambridge, Massachusetts, USA
16. *Washington University Genome Sequencing Center*, St. Louis, Missouri, USA

TWO INSTITUTIONS CREDITED WITH PROVIDING KEY COMPUTATIONAL SUPPORT AND ANALYSIS:

1. *National Center for Biotechnology Information*, National Institutes of Health, USA
2. *European Bioinformatics Institute*, Cambridge, UK

Source: National Institute of Health news release, June 26, 2000.

It should be understood that the tremendous achievements of the Human Genome Project, headed by Dr. Francis Collins (Figure 4.8), still represent only a starting point. The original work resulting from this project—the genome "map," so to speak—is admittedly very basic. Like all original maps, it is a valuable navigational tool for future explorers. As time passes, details will be continually added

FIGURE 4.8 Dr. Francis Collins head-ed the Human Genome Project and now acts as director of the National Human Genome Research Institute (NHGRI). For those who might think that such an accomplished scientist would have little time for outside interests, think again. Dr. Collins is a motorcycle enthusiast who enjoys giving musical (frequently humorous) performances, some of which have been posted on YouTube.com. He also describes himself as a "serious Christian" and has recently published *The Language of God: A Scientist Presents Evidence for Belief* (Collins, 2006).

to the map, giving us a much clearer picture of the human genome (Stein, 2004). The remainder of this chapter will examine some of the broader effects the information from the Human Genome Project is likely to have on human society.

Positive implications

It is easy to discuss the benefits of the Human Genome Project. As described in **Box 4.2**, the potential for tailoring drug therapy to specific subtypes of disorders is an excellent example. The study of how genetics affects responses to particular drugs, called pharmacogenomics, was discussed in Chapter 3. For students who enjoy the study of both pharmacology and genetics, this is an area of study worth investigating, given the seemingly limitless potential for its future expansion.

The use of genetic markers as risk factors for general debilitating conditions such as obesity, hypertension, heart disease, diabetes, and any number of other disorders will also be invaluable. Imagine, for example, a genetic screening test done in early adulthood that indicates a high risk of heart disease. An individual armed with this knowledge could improve eating habits, increase cardiovascular exercise, undertake relaxation and meditation regimens, or incorporate any number of other preventive strategies to reduce the risk of heart problems long before they occur. Keep in mind that there may be a genetic predisposition toward many disorders, including some forms of cardiovascular disease, which means that they are influenced, but not necessarily predetermined, by genetics. In the case of such disorders, knowledge is a powerful resource for establishing a wellness program designed to reduce the expression of dysfunctional features in the phenotype.

BOX 4.2   The Human Genome Project and Pharmacogenomics

When a map of the genome for "normally" functioning individuals is established, it will represent a basis of comparison to be used against samples from individuals affected by schizophrenia, depression, ADHD, or any of the hundreds (maybe thousands) of disorders that have a genetic component. More specifically, think back to the discussion in Chapter 3 about the advances that could be made, particularly in pharmacogenomics, with a detailed map of the human genome.

To put this possibility in perspective on a personal level, imagine, for example, that you begin to suffer from chronic depression. Your psychiatrist suggests treatment with an antidepressant drug. But which drug will work? A diagnosis of depression could arise from dysfunction in any of a number of neurotransmitter systems, including the serotonin, norepinephrine, dopamine, and GABA systems. Several classes of drugs, including MAO inhibitors, tricyclic antidepressants, selective serotonin reuptake inhibitors, and selective norepinephrine reuptake inhibitors, are used as treatments for depression. All of these classes of drugs alter different neurotransmitter systems or combinations of systems. In addition, there are multiple subtypes within each of these classes. Thus, there are literally dozens of possible treatments for depression. Psychiatrists have only a limited understanding of which drugs are likely to work most effectively for each individual. Furthermore, among the drugs that do work, it is diffi-cult to know in advance which will produce the fewest side effects.

Now look to the future. Imagine that the initial visit to your psychiatrist includes a simple blood test that shows a defect in your *HTR1B* gene, located at q13 on chromosome 6. A look at the database tells your psychiatrist that this gene codes for the 1B subtype of the serotonin receptor, and that it has been implicated in contributing to substance abuse disorder and major depression, but not to bipolar disorder, schizophrenia, or suicide attempts (Huang et al., 2003). Wow! This is a tremendous amount of information to use in selecting a course of treatment. In addition, when your psychiatrist establishes an effective treatment for your case of depression, that information will be added to the database for future use by researchers.

The future will hold an expansive database of the most effective treatments identified through an extensive process of screening for all possible gene combinations known to contribute to psychiatric disorders. The guesswork associated with pharmacological treatment will be vastly reduced. More effective drugs, drugs with fewer side effects, and even vaccines may be among the products of this area of study. Finally, rapid, effective treatment and disease prevention will mean lower health care costs and more productive individuals. The Human Genome Project has the potential to pay back the funds that were originally invested in it many times over.

In the case of disorders that are predetermined by genes, the ability to identify genetic aberrations is immensely helpful in determining the best course of treatment and preparing the individual for the inevitability of their phenotypic expression. For example, knowing that one parent has Huntington disease, a person can elect to be tested to determine whether they have inherited the Huntington disease allele. The chance of inheriting this deadly disorder

the disease. There are, in fact, alleles of multiple genes that have been identified as contributing to the development of this and many other diseases. In these cases, it is the unknown elements that create problems. These unknown elements may include other genes that contribute to the development of the disorder or environmental factors that are required for expression of the phenotype. Thus, it is likely that future screening tests will frequently identify Alzheimer disease alleles in people who will never experience symptoms of the disease because they lack the other essential contributing elements. Such a "false positive" diagnosis could severely affect an individual's quality of life, in a most extreme case leading to chronic anxiety or depression associated with anticipation of a debilitating disease that will never develop.

A second potential problem with genetic testing is the possibility of false positive results due to sample misidentification, sample contamination, or any number of other possibilities. This problem is not a direct negative consequence of the Human Genome Project, but with such a powerful database, there is often an assumption of similar validity in the associated tests. As with any other commodity, when genetic screening becomes widely available, there will be a wide array of vendors for this service. Some will be less responsible than others, and errors will occur. Imagine the consequences of telling individuals that they are likely to develop a deadly or debilitating disorder, when in fact there is no evidence to suggest such a possibility (Figure 4.10).

A third potential problem posed by the knowledge gained from the Human Genome Project is the dilemma it presents some people by offering them accurate predictions of their future phenotypic development. Knowing that they possess an allele for a fatal disorder may help some people prepare for their inevitable fate, but others may find the certainty of such news more difficult to live with than the uncertainty of not knowing. Consider again for a moment the consequences of the highly reliable test for identifying the Huntington disease allele. Half of those undergoing the test emerge elated with the knowledge that they are unaffected. The other half emerge with no hope for a normal life. What would you do? If one of your parents were a carrier of Huntington disease, would you be tested? Would you want to know?

Finally, there is the question of the effectiveness of gene therapy and its possible side effects. Although gene therapy has been used successfully to reverse some disorders, there have also been individuals who have developed complications associated with experimental treatments. For example, at least two individuals treated with gene therapy for SCID developed leukemia shortly after undergoing treatment (Marshall, 2003). Although the number of individuals so adversely affected is small, more research is clearly needed before gene therapy will become a safe and effective form of treatment. Until that level of knowledge is gained, gene therapy will carry with it the potential for severe, even lethal, consequences.

FIGURE 4.10 Speculation about both the positive and negative outcomes of the Human Genome Project has fueled the imaginations of fiction writers. In 1997 the movie GATTACA was one of the first commercially successful Hollywood productions based heavily on concepts of human genome manipulation. In this scene, Ethan Hawke and Uma Thurman submit to a futuristic roadside DNA test.

Legal implications

No discussion of the Human Genome Project would be complete without considering the possibility of abuse of genetic information. There are indeed some observers who were well informed about the scientific basis for the Human Genome Project, yet still opposed it. Most of these individuals had concerns not about the primary research objectives, but about how the resulting information would be handled. One such concern is the possibility of health-related genetic information being misused by insurance providers. Currently, life insurance companies may request a blood sample to prescreen clients for life-threatening infections such as hepatitis or HIV. It is reasonable that a person infected with a fatal illness could be denied a life insurance policy, or required to pay a higher premium. But what if a blood sample were screened to determine whether an individual possessed alleles associated with quantitative trait disorders? If an individual's genetic makeup suggested a predisposition to alcoholism, for example, could that individual be charged higher life insurance premiums because alcoholics have a statistically higher probability of death from automobile accidents or alcohol-related disease? If some individuals' genes predicted a greater than average chance of their developing type 2 diabetes, could the cost of their health insurance be raised, or could they be denied insurance because they are at risk of needing long-term treatment? These are just some of the questions that advancing technology will raise in the years ahead.

FIGURE 4.11 Demands for producing greater volumes of consistent results in genetic screening will require skilled individuals to use and maintain increasingly sophisticated equipment. Human error in laboratories that depend on such equipment could produce unreliable results.

Beyond these issues with insurance, imagine further how prospective employers could misuse genetic screening. Currently, employers may screen job applicants and employees for recreational drug use by testing blood, urine, or hair samples. It is reasonable that a job applicant who shows high concentrations of illegal recreational drugs could be rejected. But is it reasonable that a job candidate with an allele indicating a predisposition to drug addiction could be rejected? How about an applicant with a genetic profile associated with antisocial behavior? Employers could potentially justify discriminatory hiring practices based on a wide variety of genetic information (Figure 4.11).

Although laws are currently being enacted to protect individuals against misuse of information about their genetic makeup, the process remains incomplete. The U.S. federal government, in particular, has lagged behind individual states in ensuring the privacy of genetic information and guarding against its misuse. Even if federal laws are enacted to ensure the privacy and security of genetic information, the possibility of misuse persists. No rights can be guaranteed unconditionally, especially when technology is so readily available for violating those rights. This is indeed the potential dark side of the research findings associated with the Human Genome Project.

Keeping the facts straight

Looking back to the beginning of this section and recalling our discussion of opposition to the Human Genome Project, a couple of points are worth noting here. There is, admittedly, a link between the Human Genome Project and stem cell research, although it was not part of the project's original goals. In some cases, stem cells are used as recipients for vectors carrying new DNA to develop potentially therapeutic cell lines. Details about the use of stem cells in gene therapy are discussed in the final chapter of this text. In general, because stem cells have the potential to differentiate into many types of human tissue, they provide a potentially useful vehicle for delivery of gene therapy in a range of individuals. Researchers are also experimenting with genetically modified stem cell lines that might be useful in treating hemoglobin disorders, particularly sickle-cell anemia (Persons, 2003). There is also little question that findings from the

Human Genome Project are being used for cloning experiments in plants and animals. And, although no human has yet been cloned at this writing, it is reasonable to argue that the Human Genome Project provided information that might be useful for such work.

One needs to step back, however, and look at the bigger picture. The scope and importance of the Human Genome Project are such that its results are likely to be linked to a wide range of projects, some obviously more desirable than others. But in reality, this is the fate of useful technological advances in any field. There is no denying that advances in psychoactive drug research have spawned new forms of abuse and addiction. Similarly, developments in information sharing technology have been used for pedophilic stalking as well as widespread distribution of pornography and propaganda from hate groups. It would be difficult, however, to argue that advances in either psychoactive drug research or computer technology have harmed our society as a whole. Perhaps some of the lessons learned from these and other earlier scientific advances will help us in establishing guidelines for the use of the information now emerging from the Human Genome Project.

Chapter Summary

DNA is small, and there is no current technology available to magnify its physical image. However, DNA can be studied by producing an extremely large number of copies of a specific segment. The polymerase chain reaction (PCR) utilizes the self-replicating nature of DNA to produce a billion or more copies of a segment for analysis. This amplification process allows some physical features of DNA to be identified and analyzed. With PCR, comparisons can be made between DNA samples to establish similarities and differences. Its practical applications include establishing paternity, matching criminal suspects to DNA collected from crime scenes, and identifying genetic aberrations associated with specific diseases.

The techniques used for amplifying and analyzing DNA are conceptually simple. First, a DNA sample is obtained; then a segment of interest is selected, which must lie between two flanking regions(known DNA sequences). The DNA sample is incubated in a buffer solution that contains primers, which are short base sequences that initiate DNA replication from the flanking regions inward through the segment of interest; an abundance of the four nucleotides (A, T, C, G) required for replication; and a polymerase enzyme that catalyzes DNA replication. The buffer is repeatedly heated and cooled to initiate and terminate a rapid and continuous sequence of replication cycles. The three steps of each cycle are (1) heating to encourage denaturation, (2) cooling to encourage annealing, and (3) reheating to encourage extension. The exponential nature of DNA replication means that within 2 to 3 hours over a billion copies of DNA can be created using this elegant, yet simple, technique.

Once a DNA segment has been amplified with PCR, electrophoresis can be used to visualize features of that segment. Electrophoresis uses a positive electrical charge to draw negatively charged DNA fragments through microscopic pores in a thin block of agarose gel. Larger fragments of DNA travel through the gel more slowly, and thus move a shorter distance in a given time, than smaller fragments. Amplified samples of DNA are stained with dark or fluorescent dye and form distinct bands at different locations in the gel, indicating the relative size of the segments in the sample. Electrophoresis can be used to identify some diseases caused by alleles that result from expansions or deletions of DNA segments.

Many known segments of DNA have differences in their base sequences between individuals, called polymorphisms. Restriction enzymes can be used with PCR and electrophoresis to visualize polymorphisms in a method called DNA fingerprinting. The likelihood of two individuals yielding identical fingerprints from a single segment is very low, and the likelihood of two individuals yielding similar fingerprints from multiple segments decreases exponentially. A typical analysis used to positively identify a criminal suspect might utilize five to ten sequences, yielding a probability of less than 1 in 7 billion that two individuals will have the same fingerprint.

In the 1990s, the Human Genome Project was conceived with an overarching goal of mapping the entire human genome. The project had six primary goals: (1) identify all genes in human DNA; (2) determine the base sequence of the human genome; (3) store the information in databases; (4) continually improve tools and methods for data analysis; (5) make data and technologies available to the private sector; and (6) continually address ethical, legal, and social issues arising from the project. Originally expected to take 15 years, the project accomplished its goals several years ahead of schedule.

Among the findings of the Human Genome Project was that the human genome contains between 20,000 and 25,000 genes, far fewer than the 100,000 believed to exist when the project began. This downward revision suggests that human complexity is based in the proteome (as described in Chapter 2), rather than in the genes themselves. Another interesting implication of the reduced gene estimate is that only about 1% of human DNA represents functional genes. The importance of the remaining DNA (sometimes referred to as "junk DNA") remains a mystery.

The findings of the Human Genome Project have many useful applications, including more effective diagnosis of genetic diseases, more accurate determination of useful drug treatments, identification of genetic factors that may increase an individual's risk of common illnesses, and the advancement of gene therapy. Possible undesirable effects of the Human Genome Project are less apparent, but may include genetic testing that produces "false positive" results indicating an underlying genetic problem where none exists. Even accurate detection of genetic abnormalities may have adverse consequences for individuals who are not emotionally or psychologically prepared for such life-changing information. Clearly, researchers need to proceed cautiously in applying the findings of the Human Genome Project.

The results of the Human Genome Project also raise a number of legal and privacy issues. Some laws have already been implemented to ensure that access to genetic information is greatly restricted. One concern is that genetic information suggesting the potential for health problems could be used to deny equal access to health and life insurance policies. Similarly, employers could use genetic screening to discriminate against some job applicants.

Review Questions and Exercises

■ In addition to a sample containing DNA, list the three substances added to a buffer solution that are required for successful DNA amplification by PCR.

■ Briefly describe each of the three steps in a cycle of PCR.

■ Briefly explain the nature of the bands formed by electrophoresis and why those bands may be seen at different locations on the gel.

■ What is a restriction enzyme? How might the effects of a restriction enzyme on a particular segment of DNA vary between two different individuals?

■ Briefly explain the conceptual basis of DNA fingerprinting.

■ List the six original goals of the Human Genome Project.

■ List three useful applications of results from the Human Genome Project.

■ Briefly discuss several possible legal issues that will arise from increasingly accurate genetic screening

PART II

HERITABILITY, ENVIRONMENTAL INFLUENCES, AND METHODS OF STUDY

5

Simple
Inheritance

The previous chapters of this text have introduced an extensive range of basic information about genetics. When reading through a text that presents such a broad perspective, one way to enhance one's learning is to periodically step back to evaluate the course material that is being presented and, perhaps more important, to gather some perspective on how that material fits into the grand scheme of the text. Take a moment here to evaluate your own understanding of the material presented to this point in the text. First, you should have a conceptual understanding of the physical structure of DNA and RNA. Second, you should be able to visualize DNA interacting with RNA to influence both structure and function of the nervous system. In addition, you should have a firm grasp on the concept of genes and understand how variations in these defined segments of DNA can produce alterations in the nervous system. Finally, you should be able to make not only the direct connection between nervous system function and behavior, but also, more important, the indirect connection between genes and behavior. This is a text titled *Behavior Genetics*, but the first several chapters have overemphasized the genetics part of that title. This chapter represents a transition from that basic molecular and biological groundwork. Recall that in Chapter 2, genes were described as the fundamental units of heredity that carry information from one generation to the next. This chapter and the chapter that follows address this transmission of genetic

FIGURE 5.1 Skin color is a readily observable example of quantitative trait expression. Parents whose skin colors differ typically produce offspring with skin color in the intermediate range between their parents, representing an intermingling of gene expression.

researchers seeking treatments for these disorders. It does not necessarily mean, however, that all single-gene disorders are untreatable.

One notable exception is **phenylketonuria (PKU)**, a single-gene disorder characterized by deficiencies in the production of the enzyme phenylalanine hydroxylase (PAH). PAH is required for metabolism of phenylalanine, an amino acid that is found in many protein-rich foods. Dietary phenylalanine is used to synthesize tyrosine and several neurotransmitters. In PKU, the PAH deficiency results in an excess accumulation of phenylalanine that is subsequently converted to phenylpyruvic acid. Excessive concentrations of phenylpyruvic acid are toxic to neurons, resulting in moderate to severe mental retardation (**Figure 5.2**). Following the general rule for single-gene disorders, PKU shows a qualitative inheritance pattern, and the associated enzyme deficiency is not influenced by environmental factors. The likelihood of neurotoxicity and mental retardation, however, can be greatly reduced by altering the environment. In fact, the primary treatment for PKU is a low-phenylalanine diet, which prevents phenylpyruvic acid from building up to neurotoxic concentrations. This restricted diet is supplemented with tyrosine to compensate for the loss of natural tyrosine production caused by low phenylalanine intake. The combination of dietary restrictions and tyrosine supplements creates conditions that stimulate optimal neuronal growth and function, resulting in normal or near-normal brain development in affected individuals.

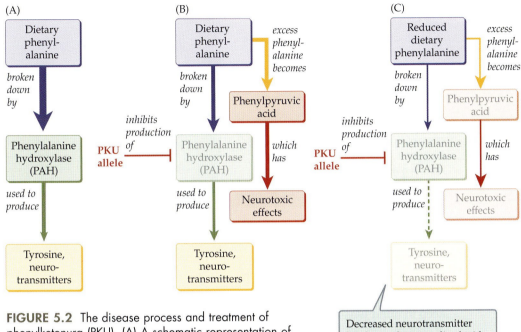

FIGURE 5.2 The disease process and treatment of phenylketonura (PKU). (A) A schematic representation of the normal metabolic pathway for dietary phenylalanine. (B) Alteration of that metabolic pathway by the PKU allele and the resulting deficiency of PAH. (C) In the presence of the PKU allele, restriction of dietary phenylalanine can reduce or eliminate the potential for PKU-induced neuronal degeneration.

Decreased neurotransmitter levels may be normalized with dietary supplements of tyrosine.

Thus, with a dietary adjustment—a simple, though perhaps not easy, environmental change (Box 5.1)—the neurotoxic effects of PKU can be nearly eliminated. Another example of treatment through environmental manipulation is a dietary change that may reduce the effects of another single gene disorder, adrenoleukodystrophy (ALD) (Moser, 2006). One interesting feature of the dietary adjustment used as treatment for this disease is that its discovery is largely credited to the parents of a child who contracted the disease. The intriguing story of Lorenzo Odone and his parents' quest to find a treatment for ALD eventually became the basis of *Lorenzo's Oil*, an entertaining and informative movie that premiered in 1992 (Figure 5.3). More information on ALD, and the use of dietary manipulation as a possible treatment, is presented in Chapter 7.

With the understanding of the distinction between qualitative and quantitative traits, the basic elements of inheritance patterns can now be explored. The remainder of this chapter focuses primarily on qualitative trait inheritance. Qualitative traits follow a simple inheritance pattern at least in terms of mathematical probabilities and compared with quantitative traits.

Mendel's Laws

Part I of this text explained the functional basis of genes. This section introduces the concepts used to understand patterns of gene inheritance as well as the influence that genes have on the development of an individual. It is interesting to note that researchers established a firm knowledge of many principles of inheritance long before understanding how genes were structured or how they functioned. Today, researchers have the luxury of an in-depth and expanding knowledge of the elements that underlie heritable traits. Our current understanding of chromosomes, DNA, RNA, and genes has helped to clarify and expand many concepts of heredity.

Gregor Mendel (1823–1884) began his first experiments with pea plants in 1856. His first published account of this work came about ten years later. This original written work describing predictable and systematic inheritance patterns went largely unnoticed, in part because his research was conducted in the Abbey of St. Thomas, outside the academic mainstream (Box 5.2). Today, however, the concepts he established nearly 150 years ago retain their theoretical usefulness and have come to be known as "Mendel's laws of inheritance."

Mendel used over 30 varieties of *Pisum sativum* (garden peas) in designing his original studies. He chose this plant species in part because of its large flowers and the wide variation in its distinguishable features, such as stems, leaves, flowers, and seeds. Perhaps more important, this plant is an **inbreeder**, meaning that it is self-fertilizing. It is also a **true breeder**, meaning that in the absence of crossbreeding or artificial fertilization, offspring will resemble their parents. These factors were crucial for Mendel's research.

Like many experiments, Mendel's original work was painfully slow. In fact, he spent the first two breeding seasons simply establishing those physical features that were clear and distinct enough to allow valid measures of change. It is worth noting that the features that Mendel selected were qualitative (stem length, for example, was defined qualitatively as "long" or "short," rather than quantitatively by measurement units). In fact, Mendel was lucky that the traits he selected were localized to single genes, making inheritance patterns highly predictable and relatively easy to observe. In addition, each of the traits selected showed a pattern of **complete dominance**, a somewhat atypical occurrence where the expression of one allele completely masks expression of the homologous, recessive allele.

BOX 5.2   The Unknown Mendel

Most readers are probably familiar with Gregor Mendel and his groundbreaking research, in which he established inheritance patterns by crossbreeding pea plants. Many readers also know that he was a monk, and that he conducted much of his original work within the grounds of a monastery. Beyond this general knowledge of his research, however, few people are familiar with the fascinating but lesser known details of Mendel's life. Mendel was born in 1822 to impoverished peasant parents in what is now the Czech Republic. By all accounts, he was a brilliant young man, and his decision to enter the monastery was probably based more on a desire to obtain a formal education (which he could not otherwise afford) than on a desire to dedicate his life to religious pursuits. Although he was eventually ordained to the priesthood, his inclination was to work as a researcher and teacher. It was in these roles that he eventually settled into a career. Although it is easy to conjure images of Mendel as a greenhouse-bound hermit, with few interests or passions outside of his work with pea plants, historians paint a much bolder picture of him. In fact, Mendel was said to have been deeply involved in social and cultural pursuits in the town of Brno, where the abbey was located. He also had research interests in meteorology and bee breeding, both of which were regular topics of courses he taught.

It is interesting to speculate what would have become of Mendel, and the fields of genetics and heredity, had he been educated in a more formal academic setting where his research pursuits would not have been focused in greenhouses and gardens. While advances in understanding of the processes of heredity were inevitable, there is no question that Mendel's work greatly advanced the theories of this discipline at a very early point in the history of the field. If you are interested in more detailed information on Mendel and his experiments, a particularly rich resource is the Web site created by the Mendel Museum located at the Augustinian Abbey in Brno, at www.mendel-museum.org.

Gregor Johann Mendel (1823–1884) is best known for his research in expression of genes and patterns of inheritance, but he also enjoyed research in meteorology and apiculture (beekeeping).

This also means that the laws of inheritance he set forth, though useful for understanding the inheritance of quantitative traits, apply directly only to qualitative traits. This point will be emphasized in greater detail in Chapter 6. For the remainder of this chapter, simply keep in mind that our discussion is directed toward qualitative traits.

The seven features Mendel ultimately selected for analysis were (1) seed color, (2) seed coat color, (3) unripe seed pod color, (4) seed shape, (5) seed pod shape, (6) flower distribution, and (7) stem length.

With these seven features established as distinct variables, Mendel began an initial series of experiments in crossbreeding that spanned seven years, from 1856 to 1863. From these experiments emerged several distinct features of inheritance and inheritance patterns. Although Mendel used the general terms "units" or "factors" to refer to what we now know as genes, his original findings translate nicely into the current terminology.

1. Individual plants possess two sets of factors, one set received from each parent. (Today we refer to these sets of factors as the genetic information making up an organism's **genotype**.)

2. These sets of factors are inherited, unaltered, from parent plants, resulting in distinctive characteristics or traits that appear in the offspring in a predictable manner. (Today these distinctive characteristics or traits are referred to as an organism's **phenotype**.)

3. Some factors dominate others and always influence the plant's characteristics, whether inherited from one parent or both. (Today we know these factors as **dominant** alleles.)

4. Some factors influence the plant's characteristics only when inherited from both parents. (Today we know these factors as **recessive** alleles.)

5. Inherited factors may be expressed or concealed, but they are never lost.

6. Any given factor is inherited independently of other factors.

In describing the characteristics of dominant and recessive factors, Mendel devised his first law of inheritance. Mendel's law of segregation states that because these paired factors are separated during meiosis, they are randomly sorted into separate gametes and randomly combined at fertilazation, ensuring variation among the offspring. Furthermore, if paired inherited factors differ, the dominant factor will be fully expressed in the organism's appearance and the recessive factor will have no noticeable effect. It is also noteworthy that although meiosis is a basic component of this law, Mendel based his original premise purely on observation of traits without an understanding of the process of meiosis (discovered some years later).

It might be useful at this point to simplify these concepts with a practical example. To do this, try to visualize some features of Mendel's experiments. Consider a pea plant that inherits from one parent a dominant allele for the seed color yellow (C). From the other parent, the plant inherits a recessive allele for the seed color green (c). That plant has a genotype of Cc for the gene for seed color and will express a phenotype of yellow seed color. Note that it was not specified *which* parent contributed the dominant and recessive alleles. That does not matter, and Mendel noted this lack of parent-specific influence on inheritance patterns in his early work. It should also be noted here that an organism with dissimilar alleles (in this case, Cc) is referred to as a **heterozygote**, whereas an organism with similar alleles (CC or cc) is referred to as a **homozygote**.

Now expand on the example. Imagine breeding a heterozygotic yellow-seeded plant (Cc) with another yellow-seeded plant possessing the same Cc genotype. What is the outcome for the offspring? Mendel's law of segregation pre-

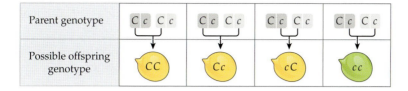

Parent genotype	C c C c	C c C c	C c C c	C c C c
Possible offspring genotype	CC	Cc	cC	cc

FIGURE 5.5 The possible outcomes of breeding two *Cc* (heterozygotic yellow-seeded) plants. In this case, the possible outcomes are one homozygotic dominant genotype (*CC*), two heterozygotic genotypes (*Cc, cC*), and one homozygotic recessive genotype (*cc*).

dicts that the possible outcomes will be two heterozygotic genotypes (*Cc, cC*) and two homozygotic genotypes (*CC, cc*) **(Figure 5.5)**.

These data can be presented simply and effectively by using a **Punnett square**, a tool developed by Reginald Punnett (1875–1967). The basic Punnett square is a simple diagram used to predict outcomes for single traits, as illustrated in **Figure 5.6**. The gametes that can be produced by each parental genotype are represented on the perimeter of the square, and possible offspring genotypes appear in the four square subunits.

The Punnett square is a simple and effective way of representing all possible genotypic outcomes, but additional interpretation is still needed. More specifically, what do these possible genotypes mean in terms of phenotypes? Carefully think this question through. Yellow seed color (*C*) is dominant, and green seed color (*c*) is recessive. In this breeding experiment, approximately three out of every four offspring (75%) will have a yellow-seeded phenotype (*CC, Cc, cC*) and one out of every four offspring (25%) would have a green-seeded phenotype (*cc*).

Take this example one step further and imagine crossbreeding two of the offspring plants. If a plant expressing the green-seeded phenotype were bred with

Cc × Cc

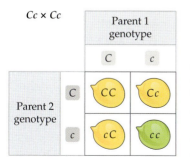

FIGURE 5.6 The Punnett square is a simple method of summarizing all possible genotypes that result from the breeding of two individuals of known genotype. The law of segregation would predict that because the two kinds of gametes are equally likely to occur, and because fertilization is random, the phenotype ratio of the offspring in this case is 3:1 yellow seeded: green seeded.

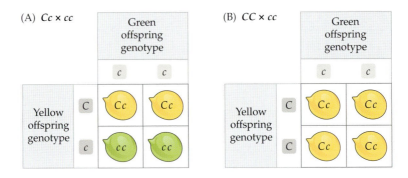

FIGURE 5.7 These Punnett squares show the possible genotypic outcomes of cross-breeding a homozygotic recessive green-seeded plant (cc) with both possible genotypes for a yellow-seeded plant (Cc and CC). (A) The phenotype ratio for green-seeded crossed with Cc yellow-seeded is 1:1 yellow-seeded:green-seeded. (B) The phenotype ratio for green-seeded crossed with CC yellow-seeded is 4:0 yellow-seeded:green-seeded.

a plant expressing the yellow-seed phenotype, what would be the outcome? This question is not complete, is it? The genotypes of the two plants being cross-bred must also be known, not just the phenotypes. As Mendel noted, genetic factors may be concealed. In this case, we know that the green-seeded plant is a recessive homozygote (cc), but what about the yellow-seeded plant? A recessive allele for the green-seeded phenotype is concealed in two of every three yellow-seeded plants (the two heterozygotes).

So, for a more complete question, if the green-seeded plant were bred with a yellow-seeded plant with a Cc genotype, what would be the expected outcome? The Punnett square in **Figure 5.7A** predicts that half of the offspring will express a green-seeded phenotype (cc, cc) and half will express a yellow-seeded phenotype (Cc, Cc). **Figure 5.7B** shows the outcome for a green-seeded plant crossed with a homozygous yellow-seeded plant (CC). In this case, all of the offspring will be heterozygotes with yellow seeds. The Punnett square makes it easy to visualize the genotypic outcome for each of these experiments.

Now, what happens if a second trait is added to the analysis. For seed shape, the dominent allele (S) is for smooth and the recessive allele (s) for wrinkled seeds. Consider then the outcomes of breeding a CcSs (yellow-seeded/smooth-seeded) plant with another plant of the same genotype. Mendel's law of segregation predicts that the seed color and seed shape alleles will segregate among the offspring. Mendel further found that the seed color alleles segregated independently of the seed shape alleles. Thus there were 16 possible outcomes: one of each of four genotypes; CCSS, ccss, ccSS, CCss, and two of each of six genotypes; CcSS, CCSs, CcSs, ccSs, CcSs, Ccss. A Punnett square can be used to visual-

Cc Ss × Cc Ss

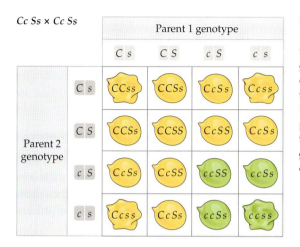

FIGURE 5.8 The basic Punnett square may be expanded to summarize possible genotypic outcomes for multiple traits. Two traits generate 16 possible equally likely outcomes. Note that individual traits generate the same ratios in 16 outcomes as they do in 4 outcomes for a single trait. For example, from the two *Cc* parents represented in this figure, the probability of a *CC* offspring genotype is 4:16 (1:4), matching the ratio of 1:4 in Figure 5.5.

ize these outcomes quickly and easily (Figure 5.8). In this case, the Punnett square is expanded from the 4 squares needed to show all possible outcomes for a single trait to the 16 squares necessary to show all possible outcomes for two traits.

This is probably a lot more information than most readers really wanted about pea plants. However, this exercise does show the usefulness of the Punnett square for assessing inheritance of multiple traits. Another important feature of this exercise is the finding that of the 16 possible phenotypic outcomes, 12 are yellow-seeded plants and 12 (though not the same 12) are smooth-seeded plants. These numbers represent the same 75% we noted when assessing the expression of the dominant color trait in the offspring of two heterozygous parents (see Figure 5.6). Traits are expressed in 25% of the cases in this two-trait example (4 of 16 are green and 4 of 16 are wrinkled).

Figure 5.8 illustrates another important point, known as **Mendel's law of independent assortment**. This law states that the inheritance and expression of one trait does not affect the probability of the inheritance and expression of other traits. In other words, it can be said that traits, in general, are inherited and expressed independently of one another.

Drawing out elaborate Punnett squares can be a bit tedious, but the usefulness of these predictable inheritance patterns, and the sheer beauty of applying simple ratios to trait expression, cannot be overstated. Students of behavior genetics must clearly understand Mendel's laws if they hope to comprehend patterns of inheritance at even the most rudimentary level. These laws have direct relevance not only for pea plants, but also for some serious human disorders (Box 5.3). You can learn more about Mendel's work, and the basic concepts of his original experiments, by visiting the Web site developed by Bill Kendrick at www. sonic.net/~nbs/projects/anthro201/. This site includes an interactive game that

FIGURE 5.9 When gametes are formed by meiosis, chromosome pairs are separated and randomly sorted into those gametes. In this case, hair color is not correlated with the occurrence of facial tics, just as Mendel found that seed color was not correlated with seed shape when he devised the law of independent assortment.

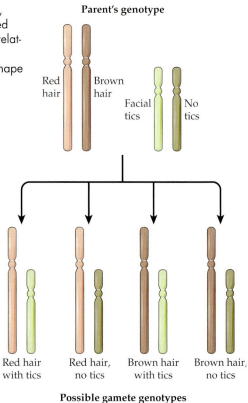

Parent's genotype

Red hair   Brown hair   Facial tics   No tics

Red hair with tics   Red hair, no tics   Brown hair with tics   Brown hair, no tics

Possible gamete genotypes

allele for each trait. As such, alleles for any two traits carried by one parent will be independently sorted into that parent's gametes, as long as those traits are carried on different chromosomes. Figure 5.9 gives an example of a physical feature (hair color) carried on one chromosome pair and a behavioral feature (facial tics) carried on a separate chromosome pair. When chromosome pairs from this individual separate to form gametes, hair color will not be correlated with facial tics. In other words, the chromosome carrying the tic allele is just as likely to be found in a gamete with a red hair allele as it is to be found in a gamete with a brown hair allele. The pea plant traits that Mendel studied followed a similar pattern of random assignment of chromosomes to gametes, allowing him to establish the basis for his law of independent assortment.

As we now know, however, some traits in humans do not seem to be transmitted independently; instead, they appear to be linked to other traits. For example, try to think of a trait that is common among individuals who have red hair. It is likely that at least one allele for red hair is linked to a separate allele that produces fair skin complexion (often with freckles). In this case, **linkage**, or expression of one phenotype that is not independent of the expression of another, violates Mendel's law of independent assortment. Linkage occurs when multiple traits are found on the same chromosome and are thus passed on as a group (Figure 5.10).

At this point, you might wonder whether the association of red hair with fair skin could result from both traits being controlled by a single gene. In other words, is it possible that there is a single gene that produces both hair color and skin complexion? If this were the case, it would resolve the problem of linkage as an exception of Mendel's law of independent assortment. Although it is possible for a single gene to influence multiple traits, a single gene is unlikely in this case, because some parents with red hair and a fair complexion produce offspring who express one or the other of these phenotypes, but not both. These cases indicate a separate genetic basis for the two related traits.

So how do alleles that are located on the same chromosome ever become separated? How does a red-haired, fair-skinned parent ever have a child with brown hair and fair skin? Such offspring are likely to be the product of **recombination**, which is a swapping of genetic material between paired chromosomes before those chromosomes are segregated in separate gametes. In simple terms, recombination occurs when chromosomes are physically pressed together during meiosis. When the paired chromosomes are in close contact they can literally swap corresponding segments of DNA (Figure 5.11). Such trading of genetic information is commonly termed **crossing over**. The result of crossing over may be the expression of a unique phenotype. Such a unique rearrangement of genetic information certainly falls outside of the simple laws of inheritance originally put forth by Mendel.

Information about linkage and recombination is also useful for mapping the locations of genes on chromosomes. Consider a hypothetical genetic disease that causes cognitive deficits. In addition, this disease often includes skin lesions and facial tics among its symptoms. Because lesions and tics are not seen in all cases, it is assumed that those symptoms are caused by defects in a linked, but separate, gene. Interesting questions arise in this case, including the question of whether lesions and tics are influenced by the same or separate genes, and how close those

FIGURE 5.10 When traits are located on the same chromosome, the law of independent assortment does not apply. Any offspring of a parent with this genotype that have red hair are likely to have a fair complexion as well, as both traits would be inherited on the same chromosome. *A*, red hair; *B*, fair complexion; *a*, brown hair; *b*, dark complexion.

FIGURE 5.11 Recombination occurs when paired chromosomes come into close contact during meiosis and exchange like segments. Each segment then fuses with the other chromosome of the pair, creating a unique combination of alleles on each chromosome. In this example, the segment of each chromosome containing the gene for complexion separates from the chromosome containing the gene for hair color. The subsequent recombination of alleles results in a chromosome that pairs red hair (*A*) with a dark complexion (*b*), and another that pairs brown hair (*a*) with a fair complexion (*B*).

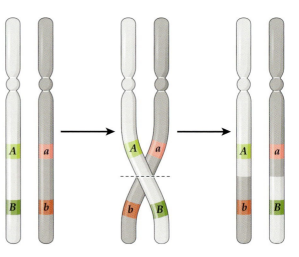

genes are to the primary gene causing the cognitive deficits. **Linkage analysis** is a method used to answer such questions. In this case, the first step would be to determine whether the lesions and tics occur independently of each other. If there are no affected individuals showing one, but not the other, symptom, it can be assumed that they are caused by defects in a single gene. If there is a difference, however—for example, if facial tics are found in 92% of affected individuals and skin lesions are found in 98%—it can be assumed not only that defects in separate genes are responsible for the two symptoms, but also that the gene associated with lesions (the symptom with the higher occurrence) is physically closer to the primary disease gene than is the gene associated with facial tics. The basis for this assumption is that genes in closer proximity on a chromosome are less likely to be separated by recombination than more distant genes (Figure 5.12).

In fact, a mathematical equation is sometimes used to quantify the distance between linked genes. More specifically, if a particular recombination occurs in 1% of gametes, then the two genes being studied are said to be 1 **centimorgan** apart. Thus, the distance between linked genes can be expressed in centimorgans based on their probability of recombination. Although the centimorgan is not a true measure of physical distance, it has been estimated that two genes recombining in 1% of gametes (1 centimorgan apart) are separated by approximately 1 million base pairs. One interesting fact about the centimorgan is that, although it is named for Thomas Hunt Morgan, the 1933 recipient of the Nobel Prize in Physiology or Medicine, it was not Morgan who devised or named the unit of measure. Rather, it was Morgan's student, Alfred Henry Sturtevant, who named the unit in his mentor's honor.

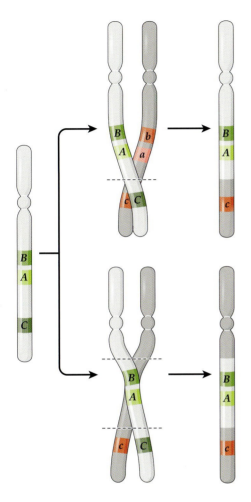

FIGURE 5.12 Linkage analysis may be used to estimate the relative locations and distances of linked genes. Consider a hypothetical disease that produces cognitive impairment (allele A). Other symptoms associated with this disease are skin lesions (allele B) and facial tics (allele C). Because of their closer proximity, it would be expected that recombination resulting in A + B would be more likely to occur than recombination resulting in A + C.

Gene conversion

When crossing over occurs during meiosis, the DNA makeup of both chromosomes is altered. In a related process of DNA exchange, the sequence of one, but not both, chromosomes is altered. When genes are transferred in this manner, the process is termed **gene conversion**. One way to conceptualize gene conversion is to envision the repair of a damaged strand of DNA that is directed by a homologous region on a paired chromosome. Although gene conversion is similar to recombination in that one chromosome assumes the information of the other, it is dissimilar in that this transfer of information is nonreciprocal (Figure 5.13).

FIGURE 5.13 Gene conversion, like recombination, involves the transfer of DNA from one chromosome to the other. It is unlike recombination in that the transfer is nonreciprocal, meaning that the donor chromosome alters the DNA sequence of the recipient chromosome, but its own DNA sequence remains unaltered.

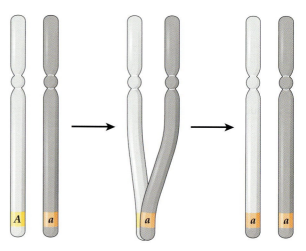

Translocation

Just as chromosomes in a pair can exchange genetic material, in some cases genetic material can be exchanged between chromosomes from different pairs (e.g., chromosome 1 may exchange information with chromosome 2). Such an exchange between nonhomologous chromosomes is called **translocation**. **Balanced translocations** are characterized by a reciprocal exchange of genetic material, whereas **unbalanced translocations** result in one chromosome gaining material while the other chromosome loses that material in the exchange process. Because unpaired chromosomes are not in close proximity as often as paired chromosomes are, translocation occurs less frequently than recombination. Mechanisms such as recombination, gene conversion, and translocation are important to understand not only because they explain variation from simple inheritance patterns, but also because they demonstrate the ability of DNA to employ alternative construction strategies. Understanding such strategies gives insight into the processes used by DNA for self-repair when damage occurs during replication.

Spontaneous mutations

Permanent, transmissible changes in genetic material are termed **mutations**. When such changes result from exposure to exogenous factors, such as viruses, chemicals, or radiation, they are referred to as **induced mutations**. Induced mutations are not considered exceptions of Mendel's laws because they do not occur as part of the natural process of genetic transmission. **Spontaneous mutations**, on the other hand, are errors in the DNA sequence that occur during replication and are perpetuated through cell division. Genetic material altered by such mutations results in phenotypes that do not occur in accordance with Mendel's laws.

Two points are worth noting here. First, although the initial occurrence of a spontaneous mutation violates the expectations of Mendelian inheritance patterns, such mutations tend to follow Mendel's predicted patterns when passed on to successive generations. Thus, Mendel's laws do hold for spontaneous mutations after they have been established in the genome. Second, the term "mutation" has such negative connotations that it is easy to believe all spontaneous mutations are detrimental to the organism. That is not the case. In fact, most spontaneous mutations are innocuous, producing only subtle changes in a species over time. This type of spontaneous mutation–induced change is consistent with the theory of gradualism. A contrasting theory of the evolutionary process proposed by Niles Eldredge and Stephen Jay Gould (1972) suggests that species evolve in phenotypic spurts, rather than by a slow, gradual process. This theory of **punctuated equilibrium** is founded in the idea that on rare occasions, spontaneous mutations occur that are important and beneficial enough to evoke rapid changes in the course of the evolution of a species (**Figure 5.14**). The theory of punctuated equilibrium is supported by many studies of fossils that show long periods of little or no change, followed by rapid changes over relatively short periods.

It is probably worth clarifying here that the theory of punctuated equilibrium is not presented as a theory opposing Darwin's concept of gradual evolution, sometimes referred to as **gradualism** when contrasted with punctuated equilibrium. Rather, the idea of rare beneficial mutations is seen as a modification to the concept of a typically gradual evolutionary progression. For example, fossil evidence of human evolution to our current physical state appears to support Darwin's implied process of gradual change. Human social and intellectual advances, however, have clearly accelerated in recent centuries, specifically since the development of language. It is conceivable that the language centers of the brain originated from a spontaneous mutation in genes that mediated the development of brain structures used for interpreting auditory information and producing vocalizations. This is a reasonable hypothesis that embellishes upon, but by no means contradicts, the theory of gradualism.

Genetic anticipation

In Chapter 4, it was noted that the basis for Huntington disease has been identified as an expansion of the DNA code at a location where a three-base sequence is repeated. The gene responsible is located in a region on chromosome 4 where the base sequence CAG may be repeated up to 35 times in an unaffected per-

The basis for genomic imprinting is unknown, but it has been the subject of some interesting speculation. Holliday (1990) suggests that a process of allelic exclusion may be a factor. Allelic exclusion occurs when one allele is active and the other is **methylated** (a methyl group is added to cytosine bases), reducing its influence during early development. Methylation may serve to reduce complications that arise when paired alleles contain conflicting instruction for a process, a situation that would otherwise result in a genetic "cross-talk" of sorts. Methylation is also known to be a host defense mechanism used to destroy the DNA of bacteria and viruses. Barlow (1993) has suggested that genomic imprinting may result from a dysfunctional defensive process, rather than a highly functional fine-tuning mechanism (as suggested by Holliday). In either case, genomic imprinting, at least in some cases, appears to result from methylation of one of two paired alleles.

Of course, any hypothesis about methylation as an underlying mechanism for genomic imprinting requires that the process differ between males and females (Figure 5.15). Such a difference has been observed in mice, in which the parent of origin influences the level of methylation of a variety of genes (Hadchouel et al., 1987; Reik et al., 1987; Sapienza et al., 1987; Swain et al., 1987).

Mitochondrial DNA

Recall from Box 2.2 that mitochondria are self-replicating. In other words, they are not produced by the cell in which they reside, but rather, arise from the division of preexisting mitochondria. It was also noted in Box 2.2 that mitochondria contain their own DNA, unique and separate from the DNA that makes up the chromosomes found in the cell nucleus. What does all of this mean in terms of inheritance patterns?

Mitochondrial inheritance

To begin with, the ground rules for mitochondrial inheritance are quite different from those of nuclear gene inheritance. Mitochondria are relatively abundant in the tail of a sperm cell, but not in the head portion that ultimately fertilizes and becomes incorporated into an egg. As a result, only maternal mitochondria are passed on to offspring. Given that, however, the inheritance of maternal mitochondria by offspring of both sexes resembles the Mendelian inheritance pattern of a single-gene dominant allele, with two notable exceptions. First, all offspring (100%) typically exhibit mitochondrially inherited traits carried by their mother. This figure contrasts with the 50% of offspring expected to be affected by traits inherited in nuclear DNA when the mother carries a single dominant allele (assuming the father does not also carry an affected al-

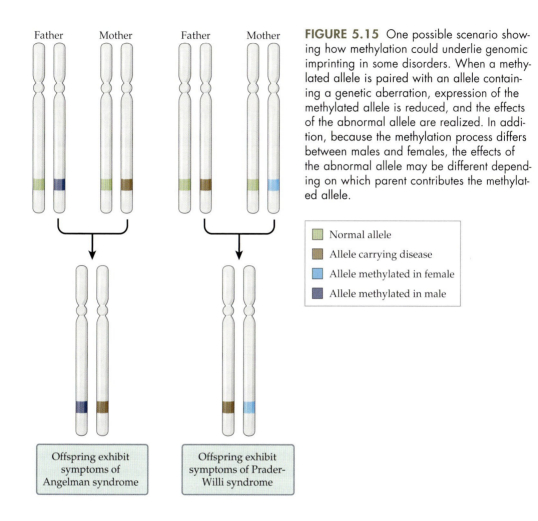

FIGURE 5.15 One possible scenario showing how methylation could underlie genomic imprinting in some disorders. When a methylated allele is paired with an allele containing a genetic aberration, expression of the methylated allele is reduced, and the effects of the abnormal allele are realized. In addition, because the methylation process differs between males and females, the effects of the abnormal allele may be different depending on which parent contributes the methylated allele.

Normal allele

Allele carrying disease

Allele methylated in female

Allele methylated in male

Offspring exhibit symptoms of Angelman syndrome

Offspring exhibit symptoms of Prader-Willi syndrome

lele). Second, mitochondrially inherited traits are never passed on through a male. In other words, males are as likely as females to be affected by a mitochondrially inherited trait, but none of their offspring will be affected. A typical inheritance pattern for a mitochondrial trait over three generations is shown in Figure 5.16.

Defective mitochondrial DNA

Diseases caused by defective mitochondrial DNA essentially result from malfunctions of the respiratory chain, which produces cellular energy by a process termed **oxidative phosphorylation**. Phenotypic features of such mitochon-

states that the inheritance and expression of one trait does not affect the probability of the inheritance and expression of other traits. The Punnett square helps us to visualize Mendel's laws and is, in general, a simple and effective way of representing inheritance patterns for dominant and recessive traits.

Although numerous exceptions of Mendel's original laws of inheritance have been noted, such special cases do not reduce the validity of those laws. In fact, they are usually explained with reference to those laws. For example, the law of independent assortment is violated when two or more traits are frequently expressed together. This phenomenon, known as linkage, occurs when two genes that produce separate traits are located on the same chromosome. Genes that are linked in this manner may be separated by a process known as crossing over, in which genetic material is literally swapped between paired chromosomes. The result of this swapping, called recombination, allows formerly linked alleles of these genes to be passed on to future generations in an independent manner.

Linkage analysis is a method of mapping the relative locations of genes on a chromosome based on their probability of recombination. Researchers have used their understanding of linkage and recombination to measure the physical distance between genes located on the same chromosome. The centimorgan is a unit of measure used to describe the distance between two genes based on their probability of recombination, with 1 centimorgan denoting a 1% probability of recombination. The physical distance represented by 1 centimorgan has been estimated at approximately 1 million base pairs.

Translocation is a phenomenon that is similar to recombination, but involves unpaired chromosomes. Balanced translocations are characterized by a reciprocal exchange of genetic material, whereas unbalanced translocations result in one chromosome gaining material while the other chromosome loses material. In gene conversion, genetic material exchange is not reciprocal (i.e., one chromosome passes genetic material to another, but does not receive any material).

Induced mutations are transmissible genetic changes that result from exposure to exogenous factors such as viruses, chemicals, or radiation. Spontaneous mutations are also genetic changes, but they do not occur as a result of any such exogenous influence. When they first occur, spontaneous mutations represent an exception of Mendel's laws of inheritance. However, spontaneous mutations that become established in the genome and are passed on to offspring follow typical patterns of inheritance as predicted by Mendel's laws.

Genetic anticipation is a phenomenon in which genetic changes of increasing severity can be observed over two or more generations. This pattern of transmitting a more apparent genetic aberration in successive generations is not adequately explained by Mendel's laws. A primary example of genetic anticipation is seen with triplet repeat expansion disorders, in which an increase in a segment of DNA is noted in unaffected parents, with an even greater increase of the same segment seen in their affected offspring.

In genomic imprinting, the phenotypic effects of an allele are dependent on the sex of the parent contributing that allele. One particular deletion from chromosome 15, for example, results in Angelman syndrome when inherited from the mother and in Prader-Willi syndrome when inherited from the father. Although these two disorders are caused by the same genetic aberration, they are characterized by markedly different physical, cognitive, and behavioral features.

Mitochondria contain their own unique DNA that is distinct from nuclear DNA in both function and inheritance patterns. One interesting feature of mitochondrial DNA is that it is passed on to offspring only from the female parent. This unique inheritance pattern means that mitochondrial genes are inherited by all offspring of a female parent, but that males are incapable of passing those traits to their offspring. Diseases caused by defects in mitochondrial DNA are expressed in tissues that are heavily dependent on oxidative phosphorylation, most notably the central nervous system and particularly the optic tracts. Other tissues that may be affected are skeletal muscles, heart muscles, kidneys, and liver. Symptoms of malfunctioning mitochondria include seizures, blindness, heart failure, and muscular dysfunction. Although the consequences of these diseases are severe, their occurrence is relatively rare.

Review Questions and Exercises

- Briefly describe the differences between qualitative traits and quantitative traits. Give an example of each.

- Single-gene disorders are generally unresponsive to treatments, but PKU and ALD represent two clear exceptions to this rule. Explain how the adverse effects of PKU or ALD may be effectively reduced.

- Define Mendel's law of segregation and Mendel's law of independent assortment.

- Create a Punnett square for a single trait showing the 4 possible outcomes. Create a second Punnett square for two independent traits, showing the 16 possible outcomes.

- Describe the phenomenon of linkage and explain how linkage analysis is used to establish the location of a gene on a chromosome.

- Define the following related phenomena: balanced translocation, unbalanced translocation, recombination, gene conversion.

- Describe genomic imprinting. Give one possible explanation for why this phenomenon occurs.

- How does the inheritance pattern of mitochondrial DNA differ from that of nuclear DNA?

Polygenic traits in animals as a model for human behaviors

The concept of polygenic traits was postulated long before current molecular techniques were available to confirm the existence of QTL. For example, over 40 years ago, Rothenbuhler (1964) suggested that the hygienic behavior of honeybees—the removal of dead pupae from their colony—was controlled by two genes (Figure 6.3). More specifically, he theorized that one gene controlled the uncapping of honeycomb cells containing dead pupae, while a distinct, separate gene controlled removal of the cell contents. Not surprisingly, researchers who recently tested Rothenbuhler's theory using modern techniques found evidence to support, and even expand upon, the original QTL hypothesis. To do this, Lapidge et al. (2002) crossbred a line of hygienic bees (which quickly and efficiently clear dead pupae) with nonhygienic bees (which slowly and inefficiently clear dead pupae). They then turned to linkage analysis (discussed in Chapter 5) to establish the locations of genes associated with hygienic behavior. Using linkage analysis, these researchers found not two, but seven, gene loci that appeared to contribute to hygienic behavior. Perhaps more interesting was evidence suggesting that each of the genes accounted for only a small percentage (9%–16%) of the variance in the behaviors that were observed for their analysis. This latter finding suggests that no single gene controls a significant

(A)   (B)

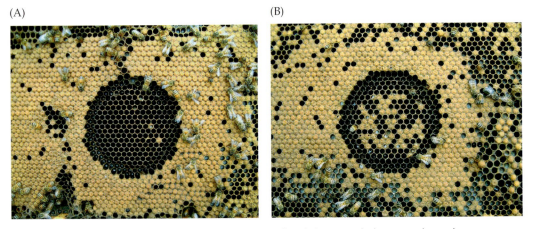

FIGURE 6.3 To study genes associated with hygienic behavior in honeybees, researchers kill pupae by freezing them with liquid nitrogen. (A) Bees bred to express extreme hygienic behavior uncapped and removed over 95% of the dead pupae within 24 hours. (B) Bees bred to express extreme nonhygienic behavior uncapped and removed a significantly smaller percentage of dead pupae within 24 hours. (After Spivak and Reuter, 2005.)

feature of hygienic behavior, but rather, that hygienic behavior is controlled by many small contributing factors.

Moving to mammals, rodent models are currently being used to expand our understanding of the influence of QTL on variable traits that are more readily applicable to humans. For example, the fear response in rodents has long been used for testing antianxiety drugs before prescribing them to humans (Box 6.1). One question that arises when using various methods of fear testing is whether all forms of fear responses originate from a single gene, or whether fear responses represent complexes of several behaviors mediated by QTL. If multiple genes are responsible, then researchers should theoretically be able to identify genetic markers associated with different types of fear responses. Recent research from the laboratory of Jonathan Flint (Fernández-Teruel et al., 2002) suggests that in the rat, chromosome 5 contains at least one gene strongly influencing generalized fear behavior. However, this extensive analysis also revealed several other genes (most notably on chromosomes 10 and 15) that contributed to other specific fear responses not influenced by the chromosome 5 gene(s). In very simple terms, the work of Fernandez-Teruel et al. strongly suggests that fear responses are behavioral traits mediated by QTL. This finding, in turn, indicates that multiple neural systems (influenced by multiple genes) are likely to be involved in most expressions of fear and anxiety.

To put this information about complex traits in perspective, try to think for a moment beyond QTL, chromosome number, and testing in rats and look at the bigger picture. The findings suggest that multiple genes contribute to multiple aspects of fear responses. In addition, some features of fear (i.e., the startle response) may have genetic influences that are distinct from those that act on other features of fear (i.e., contextual fear conditioning). Still other features of fear may share some, but not all, of their genetic influences. It is clear that fear, in rats or in humans, comes in different forms. Think of human anxiety disorders for a moment. The most recent edition of the ***Diagnostic and Statistical Manual of Mental Disorders*** (***DSM-IV***) defines 12 distinct anxiety disorders, including generalized anxiety disorder (GAD), obsessive–compulsive disorder (OCD), post-traumatic stress disorder (PTSD), and panic disorder, to name a few. Are all of these disorders manifested in the same way? Are they all treated with the same method? The answer, of course, is an unequivocal "No!"

Consider this example on a more personal level for a moment. Do you suffer from a phobia? If you do, it is probably relatively specific (heights, enclosed spaces, spiders). If you do not, it is likely that you could still be frightened by someone suddenly yelling "Boo!" in a dark, unfamiliar place. Your experience of one form of fear, but not another, probably has a basis in QTL. Fernández-Teruel and colleagues (2002) suggest that chromosome 5 may contain a gene that influences generalized fear responses in rats. A similar gene in humans could,

BOX 6.1   What Do Rats Fear?

So how exactly does a researcher test for "fear" in a rat? Remember that for rats to survive, they must avoid predators, usually by staying concealed or in the dark as much as possible. Rats are nocturnal, which gives them a natural sense of security in dark places (consider that diurnal animals, such as humans, tend to have a greater sense of fear in dark places). With this in mind, several different methods have been employed to test for fear in rat. Fear is assumed when an animal's natural tendency to explore its environment is limited. In **open field activity** tests, rats are simply placed in a large (approx. 1 meter X 1 meter) open chamber. Rats that spend more time at the periphery (along the walls) are thought to be more fearful than those that boldly venture into the central area of the chamber. Similarly, in a **light/dark box test**, rats are placed in a box with an enclosed, dark side and an open, lighted side. In this case, animals that spend more time on the dark side of the box are thought to be more fearful than those that spend time exploring the light, open side. A similar device is the **elevated plus**

maze. As the name implies, the "maze" is a four-armed apparatus (resembling a + sign) that is raised off the ground to discourage the animals from jumping out. Typically, two of the arms have walls to provide some concealment and a sense of safety from falling. The other two arms lack walls and leave the rodent exposed to more potential danger. A rat is placed in the center of the plus maze, and the amount of time it spends in the open and closed arms, and the number of movements it makes between these arms, are calculated. In this test, greater fear scores are given to animals that spend more time in the closed arms and move between arms less frequently.

In **fear conditioning**, an aversive stimulus, usually a mild foot shock, is paired with an auditory or visual cue in a novel environment. The animal responds to the aversive stimulus by "freezing." Once the animal learns to associate the cue with the aversive stimulus, it will begin to exhibit a freezing response when the cue is presented without the aversive stimulus. When the freezing response to the cue has been

in theory, mediate the sensation felt when someone yells "Boo!" Allelic variation in such a gene in humans would contribute to the variation in generalized fear seen among individuals. Specific fears such as phobias, on the other hand, may be influenced by some unique combination of the gene on chromosome 5 with other genes. In fact, findings from the rat suggest that genes influencing specific fears might be found on chromosomes 10 or 15 (Fernández-Teruel et al., 2002). Consider now how combining alleles that contribute to a common broad trait, such as generalized fear, could create a range of expression for this phenotype. Rather than resulting in expression of multiple independent fears and phobias, the combined effects of these genes would be expected to generate a continuum of expression of the broader fear trait with some elements of the fear subtypes. Such a combined effect is indeed a hallmark feature of QTL.

Recall that the hygienic behavior exhibited by honeybees is controlled by seven separate genes. Now look back at Figure 6.2 to review the number of pos-

established, subsequent tests of **contextual fear conditioning** can be performed. These tests are usually conducted 24 hours after the original fear conditioning and require only putting the animal back in the conditioning environment. Researchers then record the frequency of spontaneous freezing behavior exhibited by the animal. A particularly fearful animal will periodically freeze in the "context" of the aversive stimulus. A less fearful animal will show little or no spontaneous freezing in the same environment.

One other frequently used method of testing for fear is the **acoustic startle test**, which literally tests how much a rat jumps around when exposed to a sudden loud noise. Special chambers are used to monitor the pressure exerted by animals as they jump and move about after the startling auditory cue.

Examples of equipment commonly used to assess anxiety in rodent research. (A) An open field activity chamber. (B) A light–dark box. (C) An elevated plus maze.

(A)   (B)   (C)

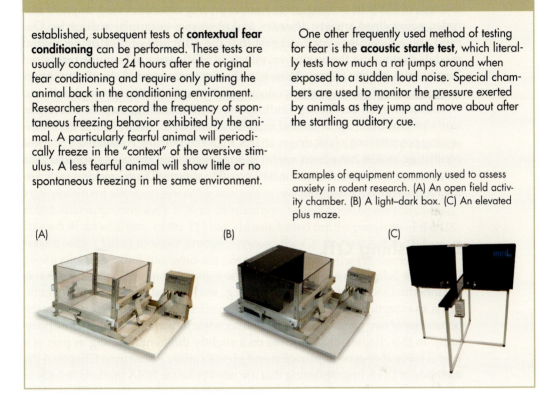

sible allele combinations that result from integrating only three genes into the expression of a phenotype. Considering all of this, it is likely that fear behaviors are mediated by far more than three genes, resulting in a wide range of types and severities for anxiety disorders in humans.

In general terms, it can be postulated that as the number of genes contributing to a trait increases, the number of possible subclasses for that trait also increases. In practical terms, this means that complex behavioral disorders may be difficult to treat, possibly requiring many different approaches based on the particular allele combinations involved in each disorder subcategory. This problem becomes clear when one considers the results of drug testing in animal models of anxiety. For example, a recent review of the literature showed that although GABA agonists and serotonin agonists show significant **anxiolytic** (antianxiety) effects on rats in open field activity tests, selective serotonin reuptake inhibitors (SSRIs) frequently do not generate significant results in these same tests (Prut

ated two important pieces of information. First, it revealed a marker that could indicate the presence of Huntington disease with a high degree of reliability (greater than 99%). Second, the finding that its predictive ability was not 100% reliable indicated that the marker itself was not part of the allele that caused the phenotype, but instead was a segment of DNA in close proximity to the allele of interest. Subsequent analyses of DNA segments in the region of the marker were ultimately used to identify the *HD* allele itself, which, as noted in previous chapters, contains an expansion of a CAG repeat from a normal range of 10–35 repeats to 37 or more. When considering the subtlety of this difference between a normal and an abnormal allele (less than 20 base pairs in some cases), it becomes apparent why the early steps in allele identification often incorporate linkage studies using markers that are easier to identify.

Population-based genetic association studies are another method commonly used to identify alleles that contribute to a phenotype of interest. Unlike family-based genetic association studies, this method requires collecting DNA samples from a large number of unrelated individuals that share a phenotype (Figure 6.5). Genetic markers that appear in a high percentage of these individuals could indicate that the markers are linked to alleles that contribute to the shared phenotype. Although the idea of scanning DNA samples for shared markers may seem simple in theory, when one considers the size of the human genome, it becomes clear that this method is impractical without some idea of where to begin the search.

FIGURE 6.5 Imagine a sample of people the size of a large crowd in which a specific phenotype is shared by everyone (e.g., everyone in the crowd has a genius-level IQ). Population-based genetic association studies assess DNA from each person in such a large sample, testing for shared genetic markers.

In some cases, known features of the phenotype provide clues to the genes that may contribute to it, offering a starting point for association studies. It is known, for example, that major depression is frequently alleviated by drugs that alter the function of the serotonin system, whereas many individuals diagnosed with anxiety disorders are effectively treated with drugs that activate the GABA system. Researchers screening large samples of individuals with these disorders would be particularly interested in genes known to influence the serotonin and GABA systems, respectively. Genes suspected of being involved in a trait are called **candidate genes**, and easily identified markers linked to these genes are among the first assessed in population-based genetic association studies. When shared genetic markers are identified among individuals with a shared phenotype, a causal relationship may then be proposed. The final step is to then determine whether the marker is a portion of the causal allele, or a closely linked sequence of DNA that does not directly influence the trait expression.

In research on human genetic disorders, a gene of interest is often referred to as a **susceptibility gene** because individuals carrying a particular allele of that gene (**susceptibility allele**) are at greater risk of developing the disease. In addition, the greater the number of susceptibility alleles a person possesses, the greater the probability of disease diagnosis. This additive effect is quite different from the Mendelian transmission of single-gene traits with qualitative and discontinuous phenotypic expression. For example, inheritance of the allele for Huntington disease (a single-gene disorder) results in a predictable pattern of motor impairment that progresses from mild to debilitating. In contrast, inherited anxiety dysfunction (a quantitative trait) may be expressed as generalized anxiety disorder, panic disorder, a phobia, obsessive–compulsive disorder, and so on. In addition, each of these possible phenotypes may be expressed over a range of severity depending on the specific combination of alleles inherited.

Basic considerations in QTL analysis

The inheritance of qualitative single-gene traits is easily predicted using Mendelian inheritance patterns. Establishing inheritance probabilities for complex quantitative traits, on the other hand, requires calculations that consider multiple genes, as depicted in Figures 6.1 and 6.2. Even so, the mathematical basis for determining QTL inheritance is fairly simple if the number of alleles underlying a trait is known. As such, it might seem that once all the genes contributing to a QTL trait or phenotype are identified, prediction of inheritance patterns would be relatively straightforward.

With QTL, however, there are several factors that reduce the value of simple mathematical probabilities of allele inheritance for predicting trait expression. One such variable is **dominance effects**, which are specific properties of a given gene as it interacts with other genes contributing to the phenotype (Box 6.2).

BOX 6.2 Dominance Deviation

The term dominance deviation is typically used to describe the relative influence of paired alleles. However, the same concept can be used to describe separate QTL genes that affect a single trait. Consider for a moment the hypothetical example of two distinct genes at separate loci. In rats, a variant of one of these hypothetical genes (allele M) results in animals that spend 20% more time in the central portion of an open field activity chamber. Inheritance of the second hypothetical gene variant (allele L) results in animals that spend 20% less time in the central portion of the chamber. In its simplest form, dominance deviation can be calculated as the difference between the mean phenotype of an animal that possesses both allele M and allele L, and the anticipated phenotype halfway between a group of animals with allele M and another group of animals with allele L. Thus, if there is no dominance effect, rats with both allele M and allele L would be expected to spend no more time in the central portion of the chamber than if they inherited neither allele. Deviation from this midpoint would suggest genetic dominance for one gene or the other—in other words, dominance deviation. For example, if an animal with both allele M and allele L spent 20% more time in the central portion of the chamber, M would be showing complete dominance. If an animal with both the M and L alleles spent 20% less time in the central portion, allele L would be showing complete dominance. The vast majority of dominance, however, is incomplete, leaving a wide range of possible phenotypes. For example, a finding that animals with both allele M and allele L spent 10% more time in the central region of the chamber would indicate partial dominance of allele M. Furthermore, because most complex traits are mediated by more than two genes, the number of variations of the phenotype is exponentially increased with each contributing gene.

Dominance effects should not be confused with the effects of a dominant allele that is expressed when paired with a recessive allele, though the general concept is similar. In the case of dominance effects, a particular allele pairing for one gene evokes a trait that masks the trait evoked by a pair of alleles for another gene that also contributes to the same phenotype. For example, if two genes contribute to premature gray hair, a person with two pairs of gray hair alleles (both genes working together) will certainly express this trait. With only a single pair of gray hair alleles (one gene contributing to gray hair) this person would express the trait only if that pair dominates the influence of the second pair. Dominance effects can be partial or complete, and they ultimately contribute to the variability of the overall joint influence of multiple genes. Typically, the greater the level of dominance for any given gene, the greater the influence of that particular allele pair on the phenotype.

Related to dominance effects is the phenomenon of **epistasis**, in which the effect of one gene is masked by the effect of a gene (or genes) that influence a separate trait. Using the example of a genetic predisposition for premature graying of hair, this trait may not be realized in an individual who also possesses a predisposition for premature baldness. In this case, the expression of one trait (baldness) does not directly affect or interact with the genetic basis for premature graying. Instead, the bald phenotype simply does not permit the graying phenotype to be realized.

Although the concept of dominance effects may seem to be a reasonable consideration in QTL analysis, an equally reasonable possibility is that genes have additive, rather than competing, effects. **Additive effects** are controlled by separate genes but combine to produce a unique phenotype or increasingly robust expression of a phenotype. As mentioned previously, Lapidge et al. (2002) found seven separate genes contributing to hygienic behavior in honeybees, each with a relatively small influence on the total behavior. This finding differed from the original hypothesis suggesting that two distinct activities associated with hygienic behavior were each controlled by single and separate genes. Instead, it is now believed none of the seven genes directly controls a distinct activity, but that an additive effect of all seven genes contributes to a single complex and variable trait.

For some behavioral disorders, dominance and additive effects can be conceptualized as culminating in a **liability threshold**. The concept of a liability threshold is fairly straightforward. As a person inherits an increasing number of alleles that contribute to a given polygenic trait, that individual nears a threshold for expression of the phenotype. Once the threshold is reached, the likelihood that the trait will be expressed increases dramatically. **Figure 6.6** depicts a plot that could apply to most disorders thought to have a QTL basis, including depression, anxiety disorders, and schizophrenia. Although the slope and height of the curve, as well as the location of the threshold (e.g. the point of clinical diagnosis), may vary, its general shape—a normal distribution curve—appropriately represents the relatively small number of individuals within the general population possessing a large number of contributing alleles and thus expressing severe (or extreme) forms of the disorder. Once again, glance back at Figure 6.2, which shows the inheritance pattern seen with QTL alleles contributing to polygenic disorders—also a normal distribution curve. If superimposed over Figure 6.6, it would show that the probability of inheriting the large number of contributing alleles (5 of 6 or 6 of 6) associated with clinical diagnosis is relatively low.

Discrepancies between normal curves for populations and for patients

As we have already seen in this chapter, the normal curve is a useful tool for describing the distribution of naturally occurring phenomena. One source of confusion, however, is the use of this curve to depict the severity of genetically in-

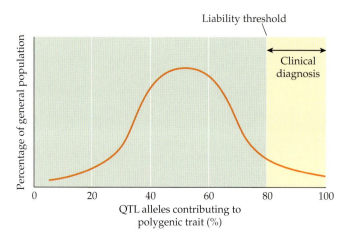

FIGURE 6.6 The concept of a liability threshold theory suggests that the number of alleles contributing to a polygenic behavioral disorder is normally distributed within the general population. In the case of anxiety, for example, few people report a complete lack of anxiety, as represented in the left tail of the curve. These individuals probably reflect the low probability of inheriting few or no susceptibility alleles. Similarly, few people report debilitating anxiety, representing those individuals who inherit a large number of susceptibility alleles, leading to a clinical diagnosis of an anxiety disorder. Finally, the greatest percentage of the population reports some nondebilitating level of anxiety, as represented by the central portion of the curve. The shape of this curve is predicted by the inheritance pattern of susceptibility alleles depicted in Figure 6.2.

fluenced behavioral disorders. More specifically, a normal distribution curve is sometimes used to show that symptoms tend to range from mild to severe, with the greatest proportion of individuals falling in the middle of the curve, reporting moderate symptoms. Such a curve is sometimes interpreted as implying that the majority of individuals in the general population suffer from moderate symptoms of a disorder.

The confusion in this case arises from the population being depicted. Typically, when researchers find a normal distribution of symptoms of a behavioral disorder within a population, the curve does not represent the general population. Instead, it represents a more specific population of interest, consisting of those individuals reporting symptoms of psychological distress. This is an important point that is worth clarification. Figure 6.6 was used to depict numbers of individuals in the general population plotted against a range of susceptibility alleles for a polygenic trait. A subsection of that figure can be used to create **Figure 6.7**, in which a normal distribution curve represents the severity of symptoms in individuals *seeking treatment*. While Figure 6.7 depicts the normal distribution of symptoms expected among individuals seeking treat-

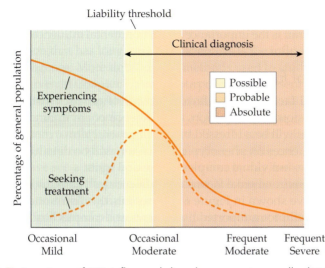

FIGURE 6.7 Symptoms of QTL-influenced disorders are not normally distributed within the general population, but rather reflect a subgroup of individuals in the right tail of that distribution. However, the distribution of symptom severity and frequency may resemble a normal distribution in the subpopulation of individuals seeking treatment. This distribution is fitted under the right tail of the normal curve that represents the number of contributing genes in the general population.

ment for behavioral disorders, perhaps the more interesting feature is the overarching tail of the larger curve representing the general population. What accounts for the difference between the two curves? It is generally accepted that not all individuals experiencing symptoms of psychiatric disorders report their experiences or seek treatment. It is also reasonable to assume that as symptoms become more severe or more frequent, the number of individuals reporting them rises closer to the actual number of individuals experiencing them. One additional assumption made when creating these theoretical curves is that the two curves will never intersect This is because a small percentage of individuals experiencing any range of symptoms will fail to ever report those symptoms.

Finally, it is worth pointing out that although the term "liability threshold" implies that when a threshold is reached, the disorder is expressed, at least two factors make such a distinction difficult in practice. First, for polygenic traits, expression is a matter of degree, rather than a distinct change. Figure 6.7 represents a hypothetical range for diagnosis of psychiatric disorders, from "possible" to "probable" to "absolute," but the distinction between these subdivisions is likely to be blurry depending on many factors. The second factor making it

FIGURE 6.9 Among the negative symptoms sometimes seen in schizophrenia are catatonic states, which are marked by the assumption of abnormal body positions that may be held for extended periods (minutes or even hours).

marked by a lack of normal behaviors,) such as social isolation, catatonic states, and blunted affect (Figure 6.9). The first class of antipsychotic drugs to be developed, sometimes referred to as **typical neuroleptics**, is characterized by relatively nonspecific antagonistic effects on all dopamine receptors. The most effective of these drugs have the greatest affinity for the D_2 dopamine receptor subtype (current theory holds that there are at least five dopamine receptor subtypes, identified as D_1–D_5). Although largely effective, these drugs produce troublesome side effects, most frequently muscle rigidity and slowing of movement referred to as **parkinsonian akinesia**.

The shortcomings of typical neuroleptics prompted further refinement of these nonspecific dopamine antagonists. With time, a new class of antipsychotic drugs, referred to as **atypical neuroleptics**, was developed. Atypical neuroleptics have a pharmacological action of blocking fewer D_2 dopamine receptors while also blocking some serotonin receptors. This newer class of drugs retains the antipsychotic properties of typical neuroleptics, but produces far fewer side effects.

Understanding at least some aspects of the neuropharmacological basis of a behavior or disorder provides a starting point for identifying candidate genes. However, as with so many aspects of behavior genetics research, relationships are seldom simple and direct. In schizophrenia research, for example, little evidence supports a role for candidate genes mediating dopamine receptor production. Linkage analysis of candidate genes for neurotransmitters that are known to mediate dopamine transmission in specific brain regions, including serotonin, GABA, and glutamate, has been more fruitful (Aghajanian and Marek, 2000; Wassef et al., 2003). These recent findings are covered in more detail in Chapter 11, but for our purposes here, imagine how multiple neurotransmitter systems might contribute to the development of a single disorder, or possibly to subtypes of a disorder. For example, inheritance of an allele that causes dysfunctional serotonin transmission may be directly associated with development of schizophrenia, but only if aberrant alleles of other genes are also present. These may include alleles that adversely affect the serotonin, GABA, or glutamate systems, or possibly other neurotransmitter systems not yet identified as being involved in the phenotype of schizophrenia.

Keeping this scenario of interactions among multiple neurotransmitter systems in mind, consider the inheritance pattern that would be expected for schizophrenia. Within the general population, there would be a low probability of inheriting the combination of contributing alleles necessary to evoke symptoms in the range of clinically diagnosed schizophrenia. Looking back at Figure 6.7, it would further be predicted that within the general population, a greater number of individuals would exhibit "mild" schizophrenic symptoms than would be diagnosed with schizophrenia. Such individuals might be those diagnosed with a range of **personality disorders** (**PD**), each of which includes relatively mild symptoms characteristic of schizophrenia. Schizotypal PD is the most relevant of these disorders, affecting about 3% of the population. The symptoms of this PD are indistinguishable from those of schizophrenia, except that they are less severe. This PD is, in fact, considered by some to be a mild form of schizophrenia. Other PDs that share a narrower range of symptoms with schizophrenia include paranoid, schizoid, antisocial, histrionic, borderline, narcissistic, dependent, avoidant, and obsessive–compulsive PDs. In 2004, Grant and colleagues estimated that over 14% of the U.S. population met the standard diagnostic criteria for at least one personality disorder.

Given that most PDs are characterized by mild forms of symptoms that are also seen in the more severe and less common disease of schizophrenia, a QTL hypothesis could be proposed. **Figure 6.10** shows a theoretical normal curve

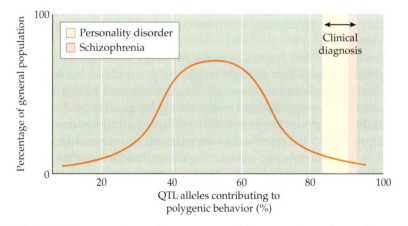

FIGURE 6.10 This curve shows the proportion of the general population that would fall into theoretical ranges for the clinical diagnosis of personality disorders and schizophrenia if it is assumed that the number of susceptibility alleles inherited is normally distributed. Such a division accurately represents prevalence reports for these disorders in the general population.

"nature-versus-nurture" debate. Although some readers may have images of a heated debate over which factor—genes or environment—is more important in influencing behavior, such divisive controversy does not exist among those who study behavior genetics. In fact, the vast majority of researchers now know that the nature-versus-nurture question must be framed within the context of specific traits, and that even within those traits; contributions of both factors are always recognized.

Evolution

In simple terms, **evolution** is a process whereby the traits of living organisms change over successive generations. Typically, these changes are **adaptive**; that is, they are changes that increase the organisms' chances of survival and reproduction. In some cases, they may even result in the emergence of new species. For those studying behavior genetics, evolution can be defined more specifically as a change in allele frequencies in a population over successive generations. **Allele frequency** refers to the relative frequency of a particular gene variant within the population. For example, a particular allele that occurs as a heterozygote in 1 out of every 10 subjects in a population would have an allele frequency of 10%. Variations of a gene that contribute to survival and reproduction would be expected to increase in frequency over successive generations.

In the previous chapters, you have learned that genes code for proteins, and that proteins are used to build every type of cell in your body. Those cells make up your entire physical being, from skin, bone, blood, and hair cells to neurons and glial cells. Thus, changes in the frequencies of alleles within a species are directly correlated with changes in its phenotype. But what gene alterations underlie these changes, and what are the processes that alter a species' genetic makeup from one generation to the next? To answer these questions, a little background in evolutionary theory is needed.

The contributions of Charles Darwin

A wide range of traits may be associated with the evolution of a species. Charles Darwin (1809–1882) was among the first to put forth a formal theory suggesting that adaptive phenotypic changes were responsible for both the persistence of species and the development of new species (Figure 7.1). Although he did not originate the concept of evolution (Box 7.1), he was the first to articulate the con-

FIGURE 7.1 Charles Darwin (1809–1882).

cept of natural selection as an explanation for evolutionary changes. **Natural selection** is the process whereby some individuals survive longer and reproduce more than other conspecifics. Furthermore, those individuals that produce the most offspring typically do so because they are the best suited to their environment. In this way, those traits most conducive to reproductive success are perpetuated in successive generations. Such traits may be as simple as coloration (i.e., better camouflage coloration may enhance survival for butterflies) or as complex as social behaviors (i.e., more elaborate pack communication behaviors may enhance survival for wolves).

In addition to proposing the theory of natural selection, Darwin also proposed a second, related theory of **sexual selection**. Like natural selection, sexual selection is a process whereby traits advantageous to reproductive success are passed on to future generations. However, whereas natural selection applies to all individuals regardless of sex, sexual selection applies to the reproductive success of individuals in relation to other individuals of the same sex.

BOX 7.1   Darwin's Contributions

Charles Darwin is among the most notable figures in the history of genetic research, and arguably in the history of all the natural sciences. Over time, however, Darwin's contributions have evolved (so to speak) to include credits he might not be due. For example, many people believe that Darwin was the first to propose the idea of evolution. In fact, the idea of evolution had been proposed and discussed in scientific circles long before Darwin published his first papers on the topic. To his credit, Darwin was the first to publish a comprehensive and cohesive work describing a theoretical basis for evolution (the process of natural selection). But the original concept appears to have been part of a debate among many scientists of his era.

Darwin is also frequently credited with coining the phrase "survival of the fittest" to describe his theory of natural selection. In fact, it was Herbert Spencer, a contemporary of Darwin, who first used this term to describe his interpretation of Darwin's theory of natural selection. Although most researchers believe the term "natural selection" is more appropriate for describing the basis for evolution, the phrase "survival of the fittest" is a popular metaphor used in many other fields. Even Spencer himself proposed this phrase as a metaphor to describe phenomena from fields as diverse as biology and economics.

Finally, although Darwin's theories of evolution now enjoy widespread acceptance and even praise from prominent scientists, this was not always the case. At the time his writings were published, many researchers actually rejected the unflattering idea that humans shared a common ancestor with all other creatures. This resistance to Darwin's theories was not surprising, considering the influence of theology on society at the time. In fact, some elements of that theological opposition to evolution are still apparent today, in spite of a preponderance of evidence supporting the theory. For readers interested in more information about Charles Darwin, resources abound, including a readily accessed Web site: www.aboutdarwin.com.

Gene–environment interaction

The relationship between environmental factors and phenotype expression is not always clear. Perhaps the clearest relationship is seen in a direct **gene–environment interaction**, wherein phenotype expression is directly altered by environmental influences. Recall from Chapter 5 the case of Lorenzo Odone, who suffered from adrenoleukodystrophy (ALD). The aberrant allele responsible for ALD has the effect of stripping the myelin from neuronal processes, literally dissolving glial cells away from axons. However, the demyelinating effect of this allele requires an accumulation of very long chain fatty acids in the body (an environmental factor). The use of Lorenzo's oil, which reduces the buildup of long-chain fatty acids, is an example of using an environmental intervention to counteract the functioning of the aberrant allele. In this case, a direct gene–environment interaction is apparent. Keep this process of gene–environment interaction in mind while reading through the following sections. Although the genetic and environmental factors may be less distinct in some examples, the basic principles of the process remain consistent throughout.

Prenatal environmental effects

In previous chapters, it was noted that some environmental factors could alter the function of genes that control the construction of proteins and cells. In this section, some of the environmental factors that may alter fetal cell development in utero are discussed.

First, consider those genetically controlled developmental processes that enhance the viability of the organism. Development of the nervous system provides a good example. It is easy to imagine genes that direct the growth and branching of axons and dendrites during fetal development. However, it is also likely that some genes influence a process of preprogrammed cell death, known as **apoptosis**. Apoptosis occurs in neurons when genes interact with specific elements of the cells' environment. For example, when axons and dendrites become part of an active communication circuitry during fetal development, the presence of neural activity appears to provide critical feedback that prevents gene-influenced apoptosis. On the other hand, neurons that do not establish functional connections fail to receive that feedback and undergo apoptosis (Figure 7.5). Under normal circumstances, apoptosis is an adaptive feature of neural development, eliminating nonessential neurons so that resources can be concentrated on active connections. In fact, the term **neural Darwinism** is sometimes used to describe aspects of this development process (Edelman, 1987).

Although neural Darwinism is normally a useful process, some environmental factors, most notably nutritional deficits and exposure to toxins, can adversely affect neural communication, leading to excessive and detrimental apop-

(A)

(B)

(C)

(D)

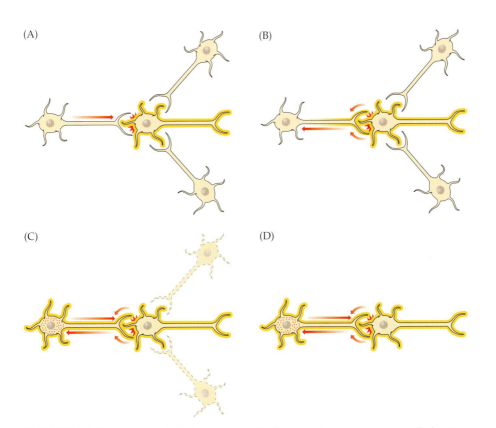

FIGURE 7.5 Programmed cell death, also called apoptosis, is a gene-controlled trait expressed predominantly during fetal development. (A) In the developing nervous system, numerous neurons may send projections to a common site. (B) Those neurons that form functional connections receive some type of feedback from the postsynaptic cell, possibly in the form of antiapoptotic chemical signals that are transported back to the soma of the presynaptic neuron. Such feedback may strengthen the connection and inhibit apoptosis. (C) Neurons that do not receive feedback succumb to genetically programmed apoptosis. (D) The end result of the apoptosis process, sometimes called neural Darwinism, is the survival of neurons that are actively integrated into the neural circuitry and the elimination of cells that contribute little or no activity to the developing nervous system.

tosis during prenatal development. This process may lead to behavioral aberrations later in life that cannot be attributed to genetic factors alone. One interesting finding is that nutritional deficiency in utero does not increase neuronal apoptosis indiscriminately. Serotonergic neurons in the hippocampal region that mediate learning and memory appear to be among the most sensitive, showing the greatest loss in nutritionally deprived fetuses (Blatt et al., 1994). This find-

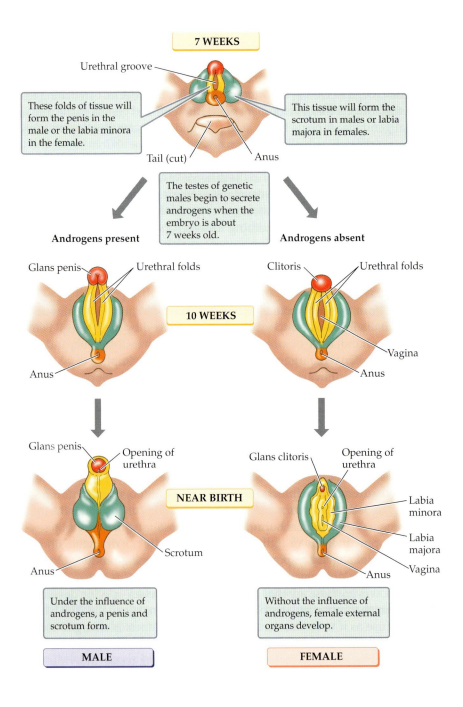

7 WEEKS

Urethral groove

These folds of tissue will form the penis in the male or the labia minora in the female.

This tissue will form the scrotum in males or labia majora in females.

Tail (cut)   Anus

The testes of genetic males begin to secrete androgens when the embryo is about 7 weeks old.

Androgens present   **Androgens absent**

Glans penis   Urethral folds   Clitoris   Urethral folds

10 WEEKS

Vagina

Anus   Anus

Glans penis   Opening of urethra   Glans clitoris   Opening of urethra

NEAR BIRTH

Labia minora

Labia majora

Scrotum

Anus   Vagina

Anus

Under the influence of androgens, a penis and scrotum form.

Without the influence of androgens, female external organs develop.

MALE   **FEMALE**

◀ **FIGURE 7.6** Testes originate from undifferentiated gonadal tissue in the abdominal cavity. Release of testosterone from the testes mediates the development of male external genitalia. Undifferentiated folds of tissue destined to become the labia majora surrounding the vaginal opening in female development instead fuse to form the scrotal sac. The final process of male genital development is the descent of the testes from the abdominal cavity into the scrotal sac.

A second disorder, known as **androgen insensitivity syndrome** (**AIS**), is perhaps even more dramatic in illustrating the effect of hormones on gene-influenced sex-specific development. AIS can result from a wide range of mutations in the *AR* gene, which codes for an androgen receptor protein (Jääskeläinen et al., 2006). The abnormal androgen receptor structure produced by these alleles does not adequately bind circulating testosterone during development. As a result, although the XY chromosomal makeup of these individuals directs the gonads to develop as testes, the testosterone released by these gonads has no masculinizing effects. In **partial AIS**, genital development takes a form similar to that in congenital adrenal hyperplasia, resulting in external genitalia with features intermediate between female and male structures. In **complete AIS**, however fully female genitalia form, making these individuals indistinguishable from XX females in appearance (Figure 7.8, Box 7.3). The significant difference between an XY AIS female and an XX female is that the AIS female has internal testes and lacks internal reproductive organs (and is therefore sterile).

Although less dramatic than the image of a genetic male with a vagina, the effects of hormones on neural development are probably more directly related to the topic of behavior. Just as external genitalia develop as fully female in individuals with complete AIS, brain structures that respond to hormones also develop as female. Like the tissues that make up the external genitalia, certain brain structures are sexually dimorphic. Among the best examples is the **medial preoptic area** (**MPOA**), a structure adjacent to the hypothalamus that plays an important role in mediating male reproductive behavior in most (if not all) mammals. The MPOA is significantly larger (containing more cells and more functional neuronal connections) in males than in females. The presence of testosterone during development probably reduces apoptosis in this region and

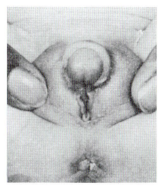

FIGURE 7.7 Adrenal hyperplasia is characterized by a pattern of genital development that is intermediate between male and female. Features include incomplete fusion of the labia, which resemble a partially developed scrotal sac, and an enlarged clitoris, which resembles a small penis.

FIGURE 7.8 Complete androgen insensitivity syndrome (AIS) results when bipotential tissues in a genetically male (XY) individual do not respond to circulating testosterone produced by the internal testes. The result is complete female development of the nervous system and of external sex-specific features (including external genitalia). In fact, the only features that distinguish an AIS female from an XX female are the presence of undescended testes in the abdominal cavity and the absence of internal female reproductive organs.

enhances dendritic and axonal connections. Female rats exposed to high levels of testosterone during fetal development show a male-like pattern of MPOA development (Sachs and Meisel, 1994). Furthermore, and perhaps most interesting, when they reach puberty, these testosterone-exposed female rats tend to show an abnormal pattern of female sexual behavior and may exhibit some male-like sexual behaviors. For example, they are reluctant to assume a receptive posture when a male rat attempts to mount them, and they may attempt to mount a sexually receptive female rat.

Based on research in rodent models, it can be assumed that human females exposed to high levels of androgens during fetal development might also display behaviors that are less feminine or more masculine than would be expected in normally developing females. The findings of one group of researchers who studied female children with congenital adrenal hyperplasia supported this theory. The researchers found that these girls tended to select more male-typical toys and engage in more displays of aggression than nonaffected age-matched girls (Berenbaum and Hines, 1992; Berenbaum and Resnick, 1997). In this case, abnormal hormone levels appear to have altered behavior patterns from those displayed when normal hormone levels interact with the influence of two X chromosomes.

It is not currently known whether the male-like behaviors seen in androgen-exposed human females are associated with structure and function changes in the MPOA. One reason for this uncertainty is that sexual dimorphism in the nervous system is not restricted to the MPOA, but is apparent in a range of neural structures and pathways, particularly in the forebrain region (Simerly, 2002). In fact, whole-brain volume is sexually dimorphic, probably reflecting the cumulative effect of a large number of individual nuclei and pathways that differ between the sexes. Most of these differences are established during fetal exposure to hormones. In this regard, hormones represent one of the most prominent factors that interact with gene function in the prenatal environment to influence behavior.

BOX 7.3   He Said, She Said: Some Thoughts on AIS

In 1985, María Patiño, a world-class female hurdler from Spain, was banned from Olympic competition when it was discovered that she was a genetic male. Physical exams for previous events in which she was allowed to compete had confirmed María's female status. It was only for Olympic competition that a blood test assessing chromosome makeup was used, but that test conclusively showed the presence of a Y chromosome. News of her male chromosome status shocked not only the sports world, but also María herself, who was

Although her genetic makeup includes a Y chromosome, María Patiño (pictured here and in Figure 7.8) considers herself no different from any other woman. The International Olympic Committee initially did not agree, but later was convinced that women with complete AIS cannot be discriminated against in Olympic competition.

unaware of her AIS condition. María later challenged the International Olympic Committee ruling that denied her eligibility, arguing convincingly that her condition did not give her an unfair advantage over other female athletes (the Olympics no longer bans individuals with AIS from competing as females).

It is now generally accepted that AIS leads to complete female development, with the exception of childbearing capability. As such, it is not surprising that many women with AIS are happily married. This leads us to an interesting consideration. The United States government has been increasingly pushing for a constitutional amendment that will ban "same-sex" marriage. Knowing what you now do about sex-specific

development, think about the irony of this for a moment. If such a ban were in place, exactly how would it define "same-sex"? If sex is defined by chromosomes, should genetic screening be required to validate marriages? What of those men who are currently married to women with AIS? Would their marriages be voided? If, on the other hand, sex were defined by external genitalia, would that mean that a man who undergoes sex reassignment surgery could legally marry another man who retained his male genitalia? One final question: Do you believe that those individuals in government who are advocating for this constitutional change have really considered all of these possible ramifications?

Postnatal environmental effects

In the previous section, you learned that many features of the prenatal environment are important in determining how genes ultimately direct the structure and function of an individual. However, as mentioned earlier, when environmental influences on behavior are discussed, most people tend to believe that the postnatal environment has the greatest effect on behavior. This is a reasonable assumption when the relative duration of exposure to the postnatal environment and to the prenatal environment (a lifetime versus a gestation period) is considered. In addition, throughout postnatal development, individuals have an ever-increasing ability to select, interact with, and respond to factors in their environment (i.e., they may actively evoke changes in their environment rather than passively responding to it). While these features do increase the potential influence of the postnatal environment, keep in mind that the influence of the prenatal environment is enhanced by chronology. In other words, because an individual is exposed to the prenatal environment first, its effects may influence how the individual responds to other factors, including the postnatal environment, for a lifetime. In short, the postnatal environment continues and expands upon the influences of the prenatal environment.

Postnatal factors that differ from prenatal factors include parenting, peer and sibling relationships, education, individual interactions with the environment, illness, injury, and so on. In fact, almost any unique experience with the postnatal environment has the potential to influence behavioral or physical development. The relative influences of many postnatal factors on development and behavior will be discussed later in this text in the context of specific phenotypes. It is sufficient to note here that within any given genotype lies the potential for tremendous variation in response to postnatal environmental factors.

While the postnatal environment offers many unique experiences that can influence gene expression, other postnatal factors are similar to those found in the prenatal environment. Just as nutrition, hormones, and exposure to toxins may affect developing fetuses, they continue to affect infants and young children, particularly in terms of neural development and behavior. Furthermore, though the types of substances, or the way in which the individual is exposed to these substances, changes after birth, their effects continue to be significant. For example, the likelihood of chronic exposure to recreational drugs (including alcohol) that is relatively high during fetal development is greatly reduced in infancy where such exposure might only occur with frequent ingestion of tainted breast milk. On the other hand, although high concentrations of lead during both fetal and early childhood development can cause brain abnormalities, prenatal exposure is uncommon compared with exposure during infancy and early childhood, when lead-containing products can be ingested by a child.

The potential for hormonal effects is reduced after birth for two reasons. First, the child is no longer exposed to the mother's hormones (or the hormones of developing siblings, in the case of multiple developing fetuses). Second, most hormones that influence development do so during certain critical periods which have passed by the end of gestation. **Critical periods** are those stages of development when environmental influences have the greatest potential to affect gene-controlled functions. Exposure to environmental influences before or after a critical period has markedly reduced or negligible effects. In essence, the critical period is a window of opportunity for the development of a structure or behavior to be altered by external factors. For example, while exposure to testosterone can have significant effects on MPOA development in a female fetus, exposure to similar levels of testosterone after birth are incapable of changing that brain structure.

Environmentally induced mutations

Up to this point, we have seen how environmental influences can alter gene function. In rare cases, environmental influences are also capable of altering gene structure. If the structure of DNA is altered by exogenous environmental factors such as radiation and various chemicals, called **mutagens**, one obvious outcome is a change in gene function. A second, rarer outcome is the passing on of environmentally induced mutations to offspring, a phenomenon that is sometimes termed **transgenerational transmission**. For example, exposure to mutagens has the potential to induce genetic changes that result in abnormal cell formation, most notably cancerous tumors. If those mutations are produced in the DNA of germ cells, they may be passed on to offspring, increasing their risk of developing specific forms of congenital cancerous cells or other hereditary anomalies (Nomura et al., 2004). Exposure to mutagens can occur during prenatal as well as postnatal development.

Gene–environment correlation

Direct causal influences of environmental factors on gene expression are fairly easy to conceptualize. As described earlier in this chapter, for example, prenatal exposure to alcohol can cause an increase in apoptosis, resulting in FAS. Other relationships between genes and environment, however, are less intuitive. **Correlational relationships** is the term applied to cases in which a significant and direct relationship exists between two or more variables, but a causal relationship cannot be established. For example, if a person reported experiencing symptoms of depression after a divorce, you probably would not assume that the divorce activated genes that then caused depression (though that could be the case). It is also possible that the person was predisposed to develop depression regardless of the divorce, but that the divorce hastened the onset of symptoms. Finally, it is possible that early symptoms of depression contributed to the divorce. In all of these scenarios, there is a clear relationship between the divorce and depression. What cannot be assumed is a direct cause-and-effect relationship between the two factors.

netics as a potential contributing factor strengthens the rationale for developing a theory that integrates both.

So, which is more important, genetic or environmental influences? The simple (yet ambiguous) answer is that it probably depends on the trait of interest. With this answer in mind, researchers generally approach the nature-versus-nurture debate by first selecting a trait and then defining the relative influences of genes and environment on that trait. When assessing the development of some traits, we find a relatively simple process with little interaction among genes. PKU, ALD, and Huntington disease, for example, are mediated by single genes. Such single-gene traits are unresponsive to most environmental factors. In terms of the nature-versus-nurture debate, these traits would be considered heavily influenced by nature. Even in these cases, however, saying that the expression of these diseases is exclusively a function of nature would be misleading when we consider that altering environmental factors can alter expression of both PKU and ALD.

More often, traits are influenced by interactions among multiple genes. Body weight, for example, is a physical feature that is influenced by more than one gene. There is probably a genetic basis for fat storage, another for metabolism, another for appetite regulation, and so on. As such, more variation in this phenotype is expected and more responsiveness to environmental factors is apparent. For example, diet and exercise can greatly influence body mass, although significant variation in body mass will be found even among individuals following very similar diets and exercise protocols.

The most complex behavioral traits are typically those that are influenced by many genes. Intelligence and personality are examples of such complex traits. Intelligence, for example, might be controlled by genes that influence short-term memory, long-term memory, object recognition, speed of processing, reading ability, spatial ability, mathematical ability, and so forth. Each of these traits, in addition to having the potential to interact with the others, could be greatly influenced by environmental factors. Such environmental influences could include learning through interactions with peers, siblings, parents, in school, through choice of career, and so forth. These potential variables and their interactions with one another would expand to create an infinite number of possible phenotypes.

Reconsidering Watson's claim

It is apparent that some traits, particularly those that are mediated by single genes, are not easily altered by environmental influences. As traits become more complex, particularly behavioral traits mediated by many genes, the potential influence of the environment increases.

In reconsidering Watson's claim—that he could use environmental influences to shape the behavior of any individual—a few points seem reasonable. First, in defense of Watson's statement, through carefully controlled environmental exposure, he probably could have created the specialists he claimed (i.e., doctor, lawyer, artist). This conclusion is based on an understanding that the traits most important for the development of those careers are complex behaviors mediated by multiple genes that are largely related to intelligence. It is probably not a coincidence that Watson did not claim he could create an accomplished weight lifter, gymnast, sprinter, or other professional athlete requiring certain physical (rather than behavioral) attributes.

Watson's claim may be overstated, however, in overlooking the reality that within any given group of individuals, there are genetic predispositions that allow some to more readily or more effectively learn particular skills than others. That is to say, any group will include some individuals whose genotypes make them better suited than other individuals for training as a doctor, a lawyer, an artist, and so on. Watson's statement is based on the idea that if the environment demands particular functions, individuals will utilize their genes to the fullest to meet those demands. For example, a person with alleles predisposing them toward artistic talent might be forged by the environment to become a doctor. In that case, the individual could use artistic skills to create diagrams explaining physiological functions. Such an individual (artist genotype, doctor phenotype) would satisfy the criteria needed to validate Watson's statement. This does not mean, however, that another individual, with a genotype that fostered inherent interest in and understanding of human physiology, would not become a better doctor under the same environmental conditions. In the same regard, we would expect that the artist alleles in a doctor phenotype might be better utilized in an environment that encouraged artistic development.

Watson, a behaviorist, certainly stimulated debate on the topic of the relative influences of environmental factors on genetic predisposition. It will be interesting to observe claims of behavior genetics researchers in the future, regarding the relative influences of genetic engineering and gene therapy. At present, many believe that manipulation of genes has the potential to dramatically alter human behavior. Perhaps the idea of changing behaviors through genetic alterations alone may prove less profound than many imagine. Indeed, in many cases it appears that the interaction between genes and the environment is indelible. The current perspective seems to be that it is an oversimplification to believe that substantial changes may be achieved by simply manipulating one variable in this equation, without regard for the other. Perhaps Watson's greatest contribution was to instill just how closely intermingled genetic influences are to environmental influences.

Chapter Summary

Evolution is a process whereby changes in allele frequencies over successive generations result in specific physical and behavioral changes in a population. Charles Darwin proposed that adaptive phenotypic changes contribute to the persistence of species as well as to the development of new species. This idea is integral to the concept of natural selection, which proposes that those organisms best adapted to the environment are most likely to survive longer and reproduce more than their conspecifics. Darwin also proposed the theory of sexual selection, which suggests that traits advantageous to the reproductive success of individuals in relation to others of their sex are passed on to future generations, regardless of their influence on basic elements of survival.

Genetic diversity within a species has important influences on the processes of natural and sexual selection. More specifically, as genetic diversity increases, so does the possibility for adaptation. Species that lack genetic diversity, and thus the ability to adapt to environmental changes, risk extinction. Adaptation to changing environments may lead to speciation, which is the evolution of a new species from a preexisting one. Adaptive radiation occurs when a single species gives rise to multiple new species, each uniquely adapted to some environmental challenge.

Inbreeding within isolated populations of a species reduces allelic variation, which can be counterproductive to the evolutionary process. However, when individuals from separate populations of the same species interbreed, allelic variation is again increased, a phenomenon known as gene flow. On rare occasions, spontaneous mutations may result in a beneficial change in a trait. Such changes in the genetic material may then become widespread in successive generations.

Although the basic concepts of natural selection apply universally across species, humans have the ability to alter some aspects of their environment to reduce the effects of potentially detrimental genes. Obesity, for example, significantly reduces the probability of reproductive success in most animals. However, because humans do not rely on strength and agility to the extent that other animals do, and because medical technology has reduced the negative effects of obesity on the process of childbirth, the effects of obesity on human reproductive success are negligible. Like genes that contribute to obesity, genes that cause dysfunctional moods and personality traits would have negative effects on reproductive success in most animals. In humans, effective psychotherapies decrease the evolutionary disadvantages of such genes.

During fetal development, a gene-controlled process of apoptosis eliminates inactive neurons. Prenatal environmental factors such as nutritional deficits can alter patterns of apoptosis, resulting in excessive or abnormal loss of neurons. Such prenatal environmental effects frequently lead to behavioral aberrations later in life that cannot be attributed to genetic factors alone.

Some human genes code for bipotential development of sexually dimorphic structures. The sex-specific development of these structures is directed by circulating hormones in the prenatal environment. Prenatal exposure to testosterone, for example, masculinizes bipotential genital tissues, which develop into male external genitalia. In some cases, bipotential tissues develop in a fashion that does not match the genetic profile of the individual. Congenital adrenal hyperplasia results when abnormally high concentrations of adrenal hormones cause genetically female individuals (XX) to develop male-like genitalia. Conversely, androgen insensitivity syndrome (AIS) causes genetically male individuals (XY) to develop female genitalia. Some brain structures are also sexually dimorphic, and those brain structures, and ultimately behavior, are affected by alterations in prenatal hormones.

The influences of the postnatal environment expand upon those of the prenatal environment. Behavioral and physical development can be influenced by a wide range of postnatal experiences, such as parenting, peer relationships, illness, and injury. Although nutritional deficits, toxins, and hormones may be factors in both the prenatal and postnatal environment, these factors are generally far less influential after critical periods of fetal development have passed. Some environmental influences can alter not only the function, but also the structure, of genes.

When genes that predispose an individual to develop a particular trait are combined with environmental factors that foster that same trait, a gene–environment correlation is established. Three types of gene–environment correlations have been proposed: (1) passive correlation, (2) reactive correlation, and (3) active correlation. Such correlations can be seen in a variety of normal traits as well as dysfunctional traits.

The "nature-versus-nurture" debate, which began more than a century ago, is based on disagreements over the relative importance of genetic makeup (nature) and personal experiences (nurture) in determining the expression of behaviors. Behaviorist John B. Watson became an iconic figure in this debate because of his extreme position that the environment determines all aspects of behavior, including character and intellect. Almost all scientists take a less polarized position, recognizing that both genes and environment play essential roles in behavioral development. However, it is generally agreed that complex traits (influenced by multiple genes) are more susceptible to environmental factors than are less complex traits (including those mediated by a single gene).

8

Research Methods
in Quantitative Genetics

Chapter 4 described some of the basic methods used to visualize DNA and explore the genome of both animals and humans. Chapters 5 and 6 explained how genes responsible for behavior are transmitted through generations. Information presented in those two chapters suggested that genetic traits follow predictable patterns of inheritance, and that those patterns may be used to predict the probability of inheriting a trait and to estimate how many genes might contribute to the trait. Admittedly, estimates of the number of genes that contribute to a trait are very rough. Chapter 7 further explored the interaction of genes and environment, with a particular emphasis on how such interactions are evident in evolution. This chapter describes research methods and designs that draw from much of the information presented in those earlier chapters. For example, modern heritability studies in animals utilize breeding protocols based on Mendel's work, while interpretations of behavioral findings depend on QTL analysis. Results from these studies may then be used to localize specific genes in animals that correspond to human genes influencing similar behaviors. Although subject to practical limitations, human studies of heritability have the potential to provide the most relevant data for determining the influences of both genes and environment on human behavior. The descriptions in this chapter of the basic methods used for human heritability studies lay the groundwork for subsequent chapters, which will revisit and expand on many of the concepts presented here.

Defining Heritability

The remaining chapters in this text focus heavily on the heritability of both normal and abnormal behavioral traits. To aid the reader in understanding the content of these upcoming chapters, a brief overview of some techniques used in heritability studies is in order. To begin with, a working definition of the term "heritability" might be useful. **Heritability** refers to the proportion of variation in trait expression that can be attributed to genetic influences and, subsequently, to the statistical probability that a particular trait will be passed on to successive generations. Methods used to estimate heritability utilize both animals and humans, typically assessing the phenotype to draw conclusions about the genotype. In some cases, these estimates can now be corroborated by data from DNA analyses using the techniques discussed in Chapter 4. In other words, once a trait has been noted in a group of individuals, researchers (in some cases) have been able to identify alleles shared by that group that may underlie the trait. The combination of data generated from heritability studies with the ability to visualize genes and their variants creates a powerful approach to predicting both the behavior of an individual and the probability of behavior expression in offspring. Before beginning a discussion of approaches used to study heritability in animals and humans, however, it is worth briefly reviewing the basic characteristics of qualitative and quantitative traits, particularly in terms of how features of these traits influence the approach taken to their study.

Qualitative traits revisited

Recall that qualitative traits are typically controlled by a single gene. In addition, some discrete form of the trait is clearly expressed in all individuals who inherit the requisite genotype. Finally, qualitative trait expression resulting from a single gene is far less likely to be affected by environmental factors than is quantitative trait expression resulting from the influence of multiple genes.

Researchers assessing the heritability of qualitative disorders frequently employ **pedigree charts**, which trace inheritance patterns over several generations. Figure 6.4, for example, showed a pedigree chart tracking the incidence of Huntington disease within an affected family. Pedigree charts are particularly useful for identifying Mendelian inheritance patterns of both dominant and recessive traits (Figure 8.1 and Box 14.1). For example, if a family history of trait expression can be traced back several generations (through medical records or personal accounts from older family members), a pedigree chart may reveal mathematical patterns that identify it as a qualitative trait. Dominant traits can be seen in successive generations, whereas recessive traits typically occur less regularly, often skipping several generations (until an unaffected carrier breeds with another unaffected carrier from outside the immediate family).

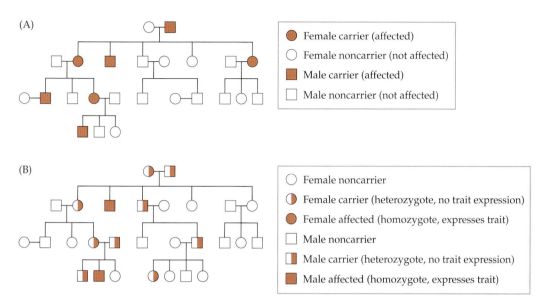

FIGURE 8.1 Pedigree charts are particularly useful for analyzing qualitative traits. (A) A hypothetical pedigree chart for a dominant qualitative trait. Historical analysis of a dominant qualitative trait will reveal expression of the trait in every previous generation. Two notable exceptions to this rule are (1) the death of an individual prior to manifestation of the trait and (2) origination of the trait as a spontaneous mutation. (B) A hypothetical pedigree chart for a recessive qualitative trait. Because recessive traits require gene contributions from both parents, trait expression will skip generations in which carriers within the family do not mate with carriers from outside of the family.

Quantitative traits revisited

Recall that quantitative traits are controlled by the joint influences of multiple genes and that they tend to be expressed along a continuum. Variation in the degree of trait expression is dependent on the relative contributions of each gene as well as on the total number of contributing alleles inherited. Furthermore, expression of quantitative traits may be affected by environmental factors. Thus, the most pronounced form of a phenotype is expected in individuals who inherit multiple contributing alleles and who are exposed to environmental influences that increase the phenotypic expression of those alleles.

Researchers assessing the heritability of quantitative traits are less inclined to use pedigree charts because these traits do not follow a simple Mendelian pattern of inheritance. In some cases, however, pedigree charts are used to show a family history of both the quantitative trait of interest and other quantitative traits thought to be related to it. **Figure 8.2** depicts a hypothetical pedigree that might be generated by researchers looking for occurrences of schizophrenia and

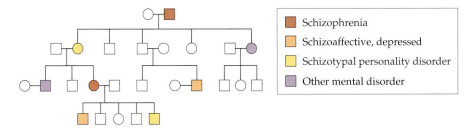

FIGURE 8.2 A hypothetical pedigree chart for a quantitative trait of interest and other, potentially related traits. Schizophrenia is known to be heritable, but pedigree charts have also revealed a higher rate of occurrence of other psychiatric disorders, including schizoaffective and schizotypal disorders, in families affected by schizophrenia than in the general population. One theory is that these related but less severe psychiatric disorders result from inheriting some, but not all, of the alleles responsible for schizophrenia.

related disorders within a family. While schizophrenia may be diagnosed in only a few individuals in a family history, other related disorders, including schizotypal personality disorder and schizoaffective disorder, may also be present, indicating a potential common genetic basis for these disorders. Although such pedigree charts are certainly less well defined than those created for qualitative traits, they do provide useful guidance in terms of traits that may contribute to, or be linked to, the phenotype of interest.

In Chapter 2, it was noted that paired chromosomes from each parent divide and are randomly distributed into gametes. Thus, parents contribute half of their genetic material to each of their offspring, which has the result that an average of 50% of genetic material is shared by any two siblings. You can easily see this pattern for yourself by doing the simple exercise in **Figure 8.3**. Like any other siblings, fraternal twins (also known as **dizygotic twins**) share an average of 50% of their genetic material. Dizygotic twins are created when two separate eggs are fertilized simultaneously. In contrast to fraternal twins, identical twins (also known as **monozygotic twins**) are created when a single fertilized egg splits soon after conception, resulting in two offspring who share nearly 100% of their genetic material.

Twin studies use these known proportions of shared genetic material to estimate genetic influences on a trait. In simple terms, greater shared expression of a trait in siblings who share 100% of their genetic material (monozygotic twins) than in siblings who share an average of 50% of their genetic material (dizygotic twins and other siblings) is assumed to indicate a genetic influence on that trait. For example, Onstad and colleagues (1991) found that if one monozygotic twin was diagnosed with schizophrenia, the second twin also met

Chromosomes	Siblings				Shared chromosomes among siblings					
	1	2	3	4	1 & 2	1 & 3	2 & 3	1 & 4	2 & 4	3 & 4
K	♦	♦	♥	♥	X					X
Q	♦	♥	♥	♥			X		X	X
J	♦	♦	♦	♦	X	X	X	X	X	X
10	♥	♥	♦	♥	X			X	X	
9	♦	♥	♦	♥		X			X	
8	♦	♥	♦	♦		X		X		X
7	♥	♥	♥	♦	X	X	X			
6	♦	♦	♥	♦	X			X	X	
5	♦	♦	♥	♦	X			X	X	
4	♥	♦	♥	♦		X			X	
3	♦	♥	♥	♦				X	X	
2	♥	♦	♥	♥		X		X		X
Ace	♦	♥	♥	♥			X		X	X
Total shared					6	6	5	7	8	6
Percentage shared					46	46	38	54	62	46

Average = 48.7%

FIGURE 8.3 To see how random distribution of chromosomes results in an average of 50% shared genetic material between siblings, try the following exercise: Take all of the red cards from a standard deck of playing cards. These will represent 13 paired chromosomes from one parent (for example, the jack of hearts and the jack of diamonds represent a homologous pair of chromosomes). Shuffle the cards to create a random assortment and then turn the cards over one at a time. Record the suit of the first card to come up for each chromosome pair (either diamond or heart) until you have recorded a suit for all 13 chromosomes. This represents the parent's random chromosome contribution to one offspring. Repeat this process for several offspring. If you create a chart similar to the one shown here, you should see an average of shared genetic material between siblings that is very close to 50%.

the diagnostic criteria for this disorder approximately 48% of the time (referred to as the **concordance rate**). In the same study, the concordance rate for schizophrenia in dizygotic twin pairs was found to be only about 4%. At a glance, the difference between the concordance rates for monozygotic twins and for dizygotic twins indicates a strong genetic influence on schizophrenia. Given that the probability that a schizophrenia diagnosis would be made in any given individual selected at random from the general population is less than 1%, the concordance rate of 4% for dizygotic twins also suggests significant heritability, further indicating a genetic basis for the trait.

A more detailed account of the methods used in twin studies, and a brief description of their mathematical basis, will be given later in this chapter. At this point, it is sufficient to understand that the heritability of quantitative traits can be evaluated even when little is known about the location or function of the contributing genes.

Animals

The ultimate goal of studying behavior genetics is to acquire an understanding of how genes, in combination with environmental factors, influence variations in human behavior. However, the logistical difficulties and ethical limitations of studying human behaviors are often highly restrictive. Thus, animal research has played a vital role in the pioneering work in the area of behavior genetics, and it continues to be one of the most fruitful current modes of heredity research. Animals can provide extremely convenient and useful models for studying patterns of gene transmission. In addition, recent findings have revealed many cases in which animal DNA contains genes that directly correspond to human genes. With all of this in mind, it is worth looking briefly at some of the many animal models used to advance the field of behavior genetics before moving on to human research.

History of animal models

When prompted to think about the study of heritability in animals, most readers probably conjure images of lab coat–clad scientists clutching clipboards and peering into cages. But imagine for a moment what the study of heritability of behavior really boils down to. It is the science of identifying the inheritance patterns of specific behavioral traits. Historically speaking, this "science" was applied to a vast array of domesticated animals long before lab coats and clipboards existed. Evidence suggests that canines were the first animals to be domesticated. Recent genetic findings indicate that the origin of all modern dogs was a single wolf species from East Asia bred by humans approximately 15,000 years ago (Savolainen et al., 2002). Think about the process that may have occurred long ago in East Asia. Humans probably began interacting with wolves (either adults or pups), selecting the most docile, social, and responsive animals from a group. Those animals were then domesticated and interbred to produce offspring that shared some of these desirable traits. From the brood, the most desirable animals were passed along to other humans who continued the interbreeding process, producing successive generations with more and more desirable traits in an early variation of the practice known as selective breeding or **trait breeding**.

Trait breeding

It is reasonable to assume that trait breeding represents one of the earliest techniques used in heritability studies. Although dog breeders from 15,000 years ago would not be considered scientific researchers, they did follow basic protocols similar to those scientific researchers use today: identify the desirable trait(s), breed animals with the desirable trait(s), generate offspring with the desirable

trait(s), repeat the previous two steps over successive generations. One remarkable outcome of trait breeding in canines is the current variety of established dog breeds, which range in size from the Chihuahua (6–9 inches, 2–6
pounds) to the mastiff (27–30 inches, 200 pounds or more). Perhaps even more
impressive than this variation in physical features is the variety of specialized
behaviors seen in particular breeds of dogs, including (but not limited to) herding livestock, killing rodents, guarding, pulling sleds, tracking animals by smell,
pointing out game animals, and retrieving (Box 8.1). The knowledge that all of

BOX 8.1   What's So Great about Labs and Goldens?

Have you ever wondered which breed of dog is
the most popular? For the past ten years or so,
the Labrador retriever has held that distinction
in the United States. How about the second
most popular breed? A close second has long
been the golden retriever. Think about these top
choices for a moment and what makes them
desirable. They are both rather large dogs with
a propensity to shed hair—not what you would
consider the most desirable physical traits in a
family pet. They are also both categorized by
the American Kennel Club as "sporting" breeds
with a history of being bred for traits required
by hunters. Are there really that many hunters
in the United States who own these breeds?
That is probably not the answer. Why, then,
have the lab and golden risen to the top of the
list of most popular breeds?

Consider what types of traits hunters desire in
a "retriever." First, they need a dog that is intelligent and can be easily trained for relatively
complex tasks. These animals are routinely
taught to follow the hand signals and whistle
commands that hunters use to direct their
retrieval of downed game. Second, retrievers
must be content to sit quietly for extended periods (in a hunting blind waiting for birds, for
example). A third requirement, and one that is
closely related to their ability to follow commands and sit quietly with hunters, is that these
animals bond closely with their owners. Finally,
retrievers are bred to have "soft mouths," a
term used for their ability to handle objects gen-

tly during retrieval. Now translate these hunting
skills to traits that are desirable in a family pet.
Retrievers are described as intelligent, with a
calm disposition, very loyal to their owners, and
exhibiting a gentle demeanor.

Interestingly, there are other breeds of retrievers that possess characteristics very similar to
those of labs and goldens but lack popularity as
pets. The flat-coated retriever and the Chesa-
peake Bay retriever remain popular among
hunters but are seldom bought as family pets.
Why these breeds have not gained greater
popularity as family pets is unclear, but
observers of dog breeding claim that this lack
of popularity has helped to maintain the integrity of their hunting skills. The logic of this theory
is that when a breed becomes more popular as
a family pet than as a sporting animal, trait
breeding for hunting features is deemphasized
in favor of trait breeding for more general com-
panion features. For example, socialization with
young children is a very desirable trait for a
family pet, whereas a highly developed sense of
smell is less important. Over generations of
breeding of these sporting dogs for primary use
as family companions, strongly social animals
might have been favored over those with the
more useful hunting trait of olfactory acuity.
Thus, the hunting integrity of the breeds might
have been compromised in this transition. The
transition itself, of course, is indicative of our
continued use of trait breeding to develop the
most desirable breeds of dogs.

(A)

(B)

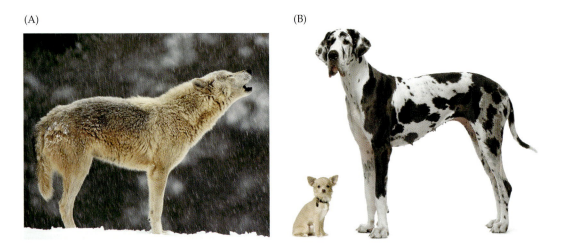

FIGURE 8.4 The ranges of both physical and behavioral traits seen among current breeds of dogs are impressive, especially considering that they are all members of the same species (*Canis lupus familiaris*). (A) All dogs are descended from a gray wolf ancestor (*Canis lupus*). (B) The American Kennel Club (AKC) has formulated seven categories of dog breeds based on both physical and behavioral traits. The difference in size between an adult Chihuahua (from the AKC "toy" category) and a Great Dane (AKC "working" category) shows the range of physical characteristics within the species (see also Figure 1.2). Note, the images are scaled to size.

these breeds can be traced to a single ancestor (the wolf) truly shows the potential for trait breeding as a means of identifying and propagating heritable traits. In addition, the fact that these specialized behaviors (as well as physical characteristics) can be successfully propagated through trait breeding over successive generations shows that these traits have a genetic basis (Figure 8.4).

Although dogs provide a popular model for describing the general features of trait breeding, the laboratory studies yielding some of the most applicable findings for our purposes have come from behaviorally less complex organisms, including bacteria, worms, insects, and rodents. The common fruit fly (*Drosophila melanogaster*) has been a key player in developing methods for studying heritability (Figure 8.5). The large body of literature devoted to fruit fly research is attributable to several features of this model organism. First, fruit flies are inexpensive, small, and hardy, which makes them excellent laboratory research subjects. Second, fruit flies have a short life cycle (about 2 weeks), which makes it easy to observe their traits over an entire life span. Third, because the fruit fly was one of the earliest model organisms in genetics and has yielded so many important findings, researchers now have a substantial database from which to work. Fourth, because the entire genome of the fruit fly has been mapped, it is possible to identify the genes associated with particular behaviors. Finally, odd as it might sound, the fruit fly has a surprisingly extensive repertoire of behaviors controlled by its relatively simple nervous system.

Early trait breeding in fruit flies focused largely on physical features. However, researchers quickly identified a number of abnormal behavioral traits in these animals, which proved more interesting fodder for heredity studies in behavior genetics. Fruit flies that express an unusual behavioral trait (sometimes referred to as **mutants**) may be bred for the trait and subjected to genetic analysis. Through this process, numerous genes have been identified that correspond

FIGURE 8.5 The common fruit fly (*Drosophila melanogaster*) has been one of the most popular animal models for the study of genetics.

to specific behavioral traits in the fruit fly. Scientists who discover a gene enjoy the privilege of giving that gene a common name. In some cases, these names are very descriptive. For example, a fruit fly that possesses an aberration of the *stuck* gene has difficulty terminating copulation. An abnormal *dunce* gene inhibits learning, while the effect of a *drop dead* gene abnormality is a very short life span, which the fly spends staggering around until it literally drops dead. Expression of a *cheap date* gene abberation is seen in fruit flies that are highly sensitive to the effects of alcohol, whereas effects of an abnormal *lush* gene are seen in fruit flies that are more attracted to alcohol than are normal flies. Expression of one allele of the *icebox* gene is seen in females that are sexually unresponsive to courting males.

Although these clever names may be humorous, keep in mind the deeper context of this information. All of the genes noted in the previous paragraph are related to behaviors that are of great interest to those researching human behavior. Research on abnormalities in learning, reproductive behavior, and responses to alcohol constitutes a large body of the medical and psychology literature. Using trait breeding to identify contributing genes in fruit flies is the first step toward understanding the basis for many human behaviors.

Inbreeding

The technique of breeding animal siblings over several generations, known as **inbreeding**, is used to produce a group of animals with a very similar, almost identical, genetic makeup. Recall that siblings share approximately 50% of their genetic material. If siblings are interbred, their offspring are certain to inherit that shared genetic material. In addition, those offspring will share approximately half of the genetic material that is not common to both of their parents (an additional 25%). Thus, first-generation inbred offspring will share about 75% of their genetic material with their siblings. If the process of inbreeding is repeated with these offspring, the next generation will inherit the shared 75% and half of the remaining 25% of genetic material that is not shared by their parents (an additional 12.5%), making these siblings approximately 87.5% similar in terms of genes. Figure 8.6 presents a graphical representation of this process repeated over six generations, at which point siblings share an average of over 99% of their genetic material—in other words, they are virtual genetic clones. Inbred lines of rodents are produced by several commercial breeders for laboratory use. These commercially available animals are typically interbred over 20 or more generations, ensuring genetic consistency when they are used for laboratory experiments.

Inbred lines of animals are useful to researchers for a number of reasons. First, when researchers assess behaviors in animals that are genetically identical, they can attribute any variation they find to environmental factors. In other words, if a group of inbred animals is randomly subdivided into separate subgroups and treated similarly, no significant differences between the subgroups should be detected. Conversely, if two subgroups subjected to different environmental factors express different behaviors, it can be assumed that difference was a result of environmental factors. Thus, inbred animals provide an ideal subject pool for assessing environmental manipulations (e.g., exposure to stress, drugs, dietary manipulations, and so on).

A second benefit is that when two or more separate lines of inbred animals are compared under the same environmental conditions, their differences can be attributed to genetic influences. One example of a study that makes good use of this method is seen in work by Cervino and colleagues (2005), who recently compared 62 strains of inbred mice in a search for genetic factors that influence plasma cholesterol levels. These researchers embarked on the laborious process of determining which of these strains had significant phenotypic differences (in plasma cholesterol levels) and comparing the genotypes of those differing stains. When they found identical genetic markers in two or more strains that exhibited significant trait differences, those regions could be excluded from their search for candidate genes (the term used for genes suspected of contributing to the phenotype). Although this particular study may not seem particularly relevant to studies of behavior, it is worth noting that cholesterol

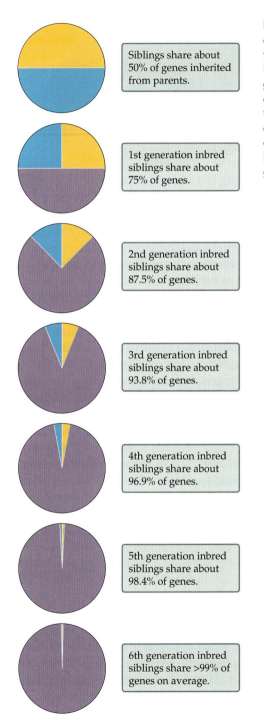

Siblings share about 50% of genes inherited from parents.

1st generation inbred siblings share about 75% of genes.

2nd generation inbred siblings share about 87.5% of genes.

3rd generation inbred siblings share about 93.8% of genes.

4th generation inbred siblings share about 96.9% of genes.

5th generation inbred siblings share about 98.4% of genes.

6th generation inbred siblings share >99% of genes on average.

FIGURE 8.6 Yellow represents inheritance of approximately 50% of genes that are not common to both parents and that are shared by a sibling. Blue represents inheritance of approximately 50% of genes that are not common to both parents and that are not shared by a sibling. Purple represents inheritance of genes that are common to both parents and thus also shared by a sibling. Within six generations of inbreeding, over 99% of genes are shared by siblings, creating animals with nearly identical genetic makeup.

recombination, gene conversion, and spontaneous mutation. Though relatively rare, a number of processes have also been found to create differences between monozygotic twin pairs by altering chromosome structure or by changing DNA sequences in individual zygotes (Gringras and Chen, 2001). In fact, recent development of techniques used to assess structural variation in DNA sequences has lead one group of researchers to suggest that small variations in DNA coding may be more prevalent in monozygotic twin pairs than current estimates suggest (Bruder et al., 2008). In this regard, while monozygotic twins certainly begin as identical genetic clones, there is a possibility of small variations in some DNA sequences as the developmental process progresses.

Some additional features of monozygotic twins are worth noting. First, about one in four pairs of monozygotic twins are **mirror twins**, meaning that distinctive features on the left side of one sibling appear on the right side in the other, and vice versa. Frequently, handedness also differs between mirror twins. In some cases, one mirror twin may exhibit **situs inversus**, a condition in which some or all of the organs are located on the opposite side of the body. The heart, for example, may be located on the right side of the chest cavity (dextrocardia) in one twin. Mirror twinning results from relatively late splitting of the zygote (typically 8–12 days after fertilization). Splitting of a zygote more than 12 days after fertilization is frequently incomplete. If such an incompletely separated zygote develops to full term, the result is conjoined twins.

In spite of these possible variations, monozygotic twins offer genetic researchers a human model roughly homologous to inbred animals. Because they share a common genetic makeup, variations in traits between monozygotic twins can be largely attributed to environmental effects. In addition, like dizygotic twins, monozygotic twins share many environmental experiences, particularly during fetal and early childhood development. For these reasons, comparisons between these two types of twins provide relevant information for studies of quantitative genetics

Consider for a moment the usefulness of twin studies designed to determine the influence of genetics on a trait. Autism is typically diagnosed during early childhood. As such, it might be reasonable to assume that prenatal environmental factors contribute to this disorder. On the other hand, the elevated rate of autism observed within some families suggests a genetic cause (Szatmari et al., 1998). Monozygotic twins are known to have a high concordance rate for autism. Specifically, when one twin is diagnosed with autism, the probability that the second twin will also be diagnosed ranges from 60% to 90% (Muhle et al., 2004). Although this finding appears to suggest a genetic basis for autism, keep in mind that monozygotic twins develop in the same environment in utero, so that exposures to chemicals, toxins, hormones, and other factors would be likely to affect both siblings nearly equally.

With all of these factors considered, the important number to evaluate here is not the concordance rate for monozygotic twins, but rather, the difference in concordance rates between monozygotic twins and dizygotic twins. If an environmental factor experienced during fetal development is responsible for autism, then concordance rates for both types of twins should be similar. In fact, the concordance rate for dizygotic twins ranges from 0% to 10%, significantly lower than the 60% to 90% seen in monozygotic twins (see Muhle et al., 2004 for review). Furthermore, the concordance rate range for dizygotic twins does not differ significantly from the concordance rate range for nontwin siblings. Taken together, these findings suggest a strong genetic influence on the development of autism. The other interesting feature of these comparison studies is that monozygotic twins do not show a concordance rate of 100%, which would indicate a single-gene disorder. Instead, the high (but not definitive) concordance rate found in monozygotic twins suggests that autism is a quantitative trait, influenced by more than one gene and with a phenotype that may be influenced by some environmental factors. Thus, the results of twin studies concur with those from pedigree studies discussed earlier. Because concordance rates are frequently used to help establish genetic influence on traits, as in this example, a few moments should be taken to understand the mathematical basis for calculating these rates (Box 8.2).

Adoption studies

You should now understand why family studies and twin studies are essential tools for determining heritability. One limitation of studies used to establish concordance rates among family members is that these studies fail to show is how much of this similarity is a function of genetic influences and how much is a function of environmental influences. Because many traits are influenced by environmental factors as well as by genetics, studies that directly assess the influence of environmental factors on the phenotype are sometimes more useful or more practical than twin and family studies. In animal research, alteration of environmental conditions, particularly during postnatal development, is easily achieved by separating siblings at a young age and raising them under different conditions. Manipulating the environment of developing children is far more difficult, particularly in terms of the ethical concerns it would raise. Thus, researchers have learned to use the naturally occurring environmental manipulations created by the adoption of children to assess behavioral traits.

Two basic methods are used for adoption studies. The first method compares the adopted individual with first-degree relatives who live in a separate environment. For example, a child's phenotype may be compared with that of the

BOX 8.2 Calculating Concordance Rates

In the upcoming chapters, you will be reading a lot about concordance rates and their use in establishing the relative influences of genes and environment on trait expression. Learning how to calculate concordance rates was probably not among your primary objectives when choosing to read this text, but knowing how these rates are determined will help you to better grasp the concepts of heritability. In fact, the mathematical formula used to determine concordance rates is very simple:

$$\text{concordance rate} = \left[\frac{\text{both affected}}{\text{(one affected + both affected)}}\right] \times 100$$

A simple example will show the practical application of this formula. Inheritance of Huntington disease (a single-gene, dominantly inherited disorder) follows a very predictable mathematical pattern. For example, if we assessed 40 monozygotic twin pairs in which one twin was affected, the trait would be found in both twins in all 40 pairs. Thus, the concordance rate would be calculated as

$$\frac{40}{(0 + 40)} \times 100 = 100$$

Indeed, research findings independently suggest a 100% concordance rate for Huntington disease among monozygotic twins (see Friedman et al., 2005). If 40 dizygotic twin pairs were assessed, the disease would occur in both twins in only about one-half of the pairs. Thus, the concordance rate would be calculated as

$$\frac{20}{(20 + 20)} \times 100 = 50$$

In dizygotic twins, the concordance rate of Huntington disease is indeed observed to be about 50%.

Concordance rates are also used to describe traits controlled by multiple genes and influenced by the environment. Major depression, for example, has been found to have a concordance rate of about 50% in monozygotic twins and about 20% in dizygotic twins. In other words, in a study assessing 40 pairs of monozygotic twins, both twins would exhibit this disorder in about 20 cases. Likewise, in a sample of 40 pairs of dizygotic twins, both twins would exhibit major depression in about 8 cases.

Finally, consider the concordance rates for **Parkinson disease**, a neurological disorder characterized by the degeneration of a specific dopamine system in the brain and a slowing of motor function. Like Huntington disease, Parkinson disease causes marked cell loss in brain regions that control movement. However, unlike Huntington disease, which shows a 100% concordance rate in monozygotic twins (indicative of a single-gene, dominantly inherited disorder), one recent study reported a concordance rate for Parkinson disease of about 11% in monozygotic twins (Wirdefeldt, 2004). In addition, that 11% concordance rate was not significantly different from the 8% concordance rate found for dizygotic twins. In this case, genes seem to contribute little to the development of Parkinson disease, which appears to be caused largely by environmental factors.

biological parents, or with that of siblings who are raised by the biological parents or by another adoptive family. In this case, observed similarities in trait expression are assumed to result largely from genetic influences (Figure 8.10). The second method compares the adopted individual with genetically unrelated individuals living in the same household (either adoptive parents or other chil-

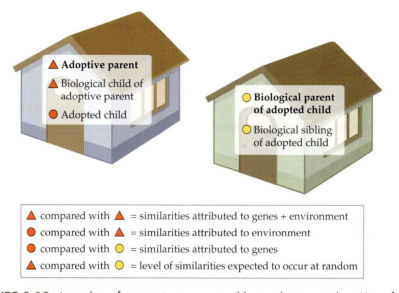

▲ compared with ▲ = similarities attributed to genes + environment
● compared with ▲ = similarities attributed to environment
● compared with ○ = similarities attributed to genes
▲ compared with ○ = level of similarities expected to occur at random

FIGURE 8.10 A number of comparisons are possible in adoption studies. Here, first-degree relatives are represented by symbols of the same shape, and individuals living in the same environment are represented by symbols of the same color. Comparing first-degree relatives living in the same environment (same shape and same color) shows similarities in trait expression influenced by shared genes plus shared environment. Comparing unrelated individuals living in the same environment (same color, different shape) shows similarities that can be attributed to environmental influences. Comparing first-degree relatives living in different environments (same shape, different color) shows similarities that can be attributed to genetic influences. Each of these comparisons can be contrasted with the degree of similarity between unrelated individuals who share no common environmental influences (different shape and different color).

dren being raised by the adoptive parents). In this case, observed similarities are assumed to result largely from environmental influences.

To better understand how adoption studies are developed, consider the simple example of a trait that is unaffected by the environment. You have learned that Huntington disease, a single-gene, dominantly inherited disorder, affects approximately 50% of the offspring of a single affected parent. If you assessed the fates of 100 adopted children, each of whom had been born to a parent with Huntington disease, you would find that about 50% of those children would develop the disease. Likewise, if you assessed 100 adopted children born to unaffected parents, but adopted and raised by parents with the disease-causing allele of the *HD* gene, you would find that none of those children would develop the disease. In this case, adoption studies would reveal no environmental influence on development of Huntington disease.

mental influences (e.g., advertising, cultural differences in diet) might be less likely to be risk factors for obesity than are physiological factors (e.g., metabolism, fat storage regulation).

Chapter Summary

Heritability refers to the proportion of variation in trait expression that can be attributed to genetic influences and, subsequently, to the probability that a particular trait will be passed on to successive generations. Heritability studies typically assess the phenotype of an organism to draw conclusions about the genotype. Qualitative traits are usually controlled by a single gene, lead to the expression of a discrete phenotype, and are influenced little by environmental factors. Researchers assessing the heritability of qualitative traits frequently employ pedigree charts, which are useful for identifying Mendelian inheritance patterns of both dominant and recessive genes.

In contrast to qualitative traits, quantitative traits are influenced by multiple genes and have phenotypes that tend to be expressed along a continuum. The degree of trait expression is often influenced by the number of contributing alleles as well as by environmental factors. Researchers assessing the heritability of quantitative traits are less inclined to use pedigree charts because these traits do not follow simple Mendelian patterns of inheritance. Instead, quantitative trait researchers calculate concordance rates, which measure the probability of trait expression in related individuals.

Animal research is an essential component of behavior genetics research. Trait breeding is a method that consists of identifying a trait of interest and then selectively breeding animals that exhibit that trait over multiple generations. The domestication of dogs represents one of the earliest uses of trait breeding. In dogs, trait breeding has resulted in a wide variety of breeds with distinctive physical features and behavioral characteristics. Whereas dog breeding represents a historical example, trait breeding of the common fruit fly has been used more recently in advancing in the study of heredity. Researchers have identified a number of genes underlying behavioral traits in mutant fruit fly models that are particularly applicable to the study of humans, including genes that control cognitive abilities, sexual behavior, and alcohol consumption.

Inbreeding is the technique of breeding animal siblings over several generations to produce a group of animals with a nearly identical genetic makeup. When the behavior of genetically identical animals varies significantly, an environmental influence on the behavior can be assumed. Conversely, when an environmental factor is held constant for separate groups of inbred animals, behavioral differences between those groups can be attributed to genetic influences. Candidate genes for a trait can be identified by finding genetic differences between inbred animals that express the trait and those that do not. Additionally, when an animal within an inbred group displays a trait anomaly, trait breeding can be used to produce a unique subset of animals that differ from the original group in terms of both trait expression and underlying genetic factor(s).

Transgenic animals have a genome that has been artificially altered, usually during fetal development. Such an alteration is typically accomplished by either inserting a segment of DNA into the existing genome (knock-in model) or destroying a segment of DNA in the existing genome (knock-out model). Transgenic animals can be used to test the effects of specific DNA sequences on specific traits.

Compared with animal research, the use of human subjects for assessing heritability has limitations, most notably the lack of inbred lines and the inability to raise and maintain subjects in a controlled environment. Researchers can estimate the degree of genetic influence on a given trait by comparing family members. First-degree relatives share about 50% of their genetic material, second-degree relatives about 25%, and third-degree relatives about 12.5%. Knowing these percentages of genetic material shared among relatives allows researchers to readily identify dominant or recessive single-gene traits. Although the results are less definitive, assessments of the heritability of quantitative traits among relatives also have research value.

Human monozygotic twins offer researchers a model wherein environmental effects on nearly identical genomes can be assessed. Because they share a common genetic makeup, variations in traits between monozygotic twins are largely attributed to environmental effects. In addition, because both monozygotic twins and dizygotic twins (who share only about 50% of their genetic material) experience a similar degree of shared environment, comparisons between these two types of twins are especially useful for establishing genetic effects in heritability studies. Most twin studies use concordance rates to quantify the relative influence of genes on a particular trait. Specifically, when monozygotic twin pairs show a concordance rate that is significantly higher than dizygotic twin pairs, this difference is assumed to be largely a result of genetic influences.

Studies of adopted individuals allow researchers to further delineate effects of genes and environment. When siblings raised in different environments are compared, observed similarities in trait expression are assumed to result from relatively pure genetic influences. When unrelated individuals raised in the same environment are compared, observed similarities are assumed to result from environmental influences.

Review Questions and Exercises

- Describe the basic process of trait breeding.

- Summarize the general uses of inbred strains of animals for behavior genetics research.

- Compare the advantages and disadvantages of using animal models and human models for behavior genetics research.

- List several reasons why fruit flies persist as a popular animal model for behavior genetics researchers.

- Describe how concordance rates are calculated.

- Create a chart that includes first-, second-, and third-degree relatives. Include the percentage of genes each relative shares with the proband.

- Briefly explain what types of comparisons can be made in adoption studies, and what kinds of factors (i.e., genetic or environmental) are viewed as responsible for the concordance rates observed in each comparison.

- Create a short list of factors that would be considered "shared environment" and a second list of factors that would be considered "nonshared environment" for the purpose of behavior genetics research.

PART III
GENETIC INFLUENCES ON BEHAVIOR AND BEHAVIORAL DISORDERS

FIGURE 9.3 It is widely accepted that several separate cognitive abilities are indicative of a more broadly defined trait referred to as intelligence. It is also generally agreed that a correlation is seen among many cognitive abilities. Based on these two observations, Spearman proposed that some core element of general intelligence (g) contributes to all cognitive functions. This schematic represents the influence of g on six cognitive abilities. The shaded area within each oval represents variance created by g, whereas the unshaded area represents variance created by the individual skills. As the diagram shows, g is not believed to affect all cognitive abilities equally. However, when the influence of g increases or decreases (imagine the gray circle growing larger or smaller), it does so with relative consistency across all abilities, accounting for the positive correlation between them. (After Jensen, 1980.)

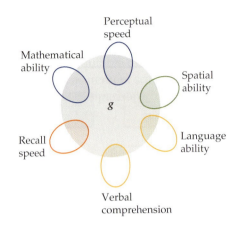

proaches used to measure intelligence in at least two ways. First, they suggest that weighted scores from multiple cognitive tests (such as IQ scores) may be used as a reasonable measure of g. This measure can then be assessed with both genetic analyses and heritability studies to establish genetic contributions to intelligence. Second, they suggest that individual cognitive abilities can be assessed independently of g to establish genetic contributions to each of these traits.

Breeding for intelligence in animals

Studies of intelligence using animal models are based on the premise that tasks designed to assess cognitive abilities may be used to measure intelligence. It is likely that trait breeding for intelligence, as mentioned in the previous chapter, began early in the history of the domestication of dogs. However, formal laboratory studies of animal intelligence began much more recently, in the early 1900s. By the mid-1900s, several researchers had found that they could breed rodents that performed efficiently (maze-bright rodents) or poorly (maze-dull rodents) in several different maze-learning tasks (Heron, 1935; Tryon, 1940; Thompson, 1954). Because each of these tasks quantified certain cognitive abilities (e.g., recall speed, spatial ability), it was concluded that proficiency at these tasks could be used as reasonable measures of intelligence.

The breeding of maze-bright and maze-dull rodents led to several revelations about the heritability of intelligence. First, the ability to breed separate lines of animals that exhibited higher or lower scores for a quantitative measure of intelligence (maze-learning ability) showed that some elements of intelligence

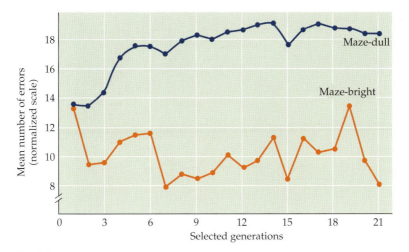

FIGURE 9.4 Researchers selected rats for inbreeding based on the number of errors they made in a maze-learning task. By breeding the animals in each generation that made the fewest errors in the maze, they established a line of maze-bright rats. Similarly, breeding the animals that made the most errors produced a line of maze-dull rats. (After Tryon, 1940.)

were indeed genetic. Second, because numerous generations of inbreeding were required to achieve significant differences in these scores, it was determined that maze-learning ability was a polygenic trait and not controlled by a single gene (Figure 9.4). Third, further evidence that maze-learning ability was a polygenic trait came from a later finding that environmental changes could significantly affect maze learning in both maze-bright and maze-dull rats (Cooper and Zubek, 1958). More specifically, when both types of rats were raised in an **impoverished environment** (deprived of movable or audible objects and lacking color variation as well as variation in object positions), maze-bright animals performed no better than maze-dull animals. And when both types of rats were raised in an **enriched environment** (provided with movable and audible objects and variation in colors and object positions), maze-bright rats performed only slightly better than maze-dull rats (Figure 9.5).

With this evidence of a hereditary factor contributing to intelligence, the next generation of researchers began searching for specific genes that might contribute to this trait. As mentioned in Chapter 8, reports of numerous genes that contributed to learning and memory in studies of fruit flies began to appear in the scientific literature during the late 1970s. As their names would imply, abnormalities of the genes *dunce*, *amnesiac*, *turnip*, and *rutabaga* each resulted in impaired learning. The identification of these genes and the mapping of their in-

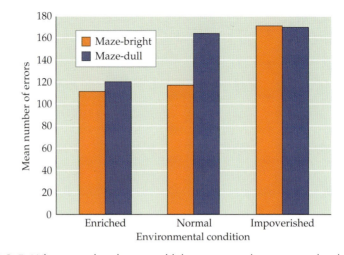

FIGURE 9.5 When raised under normal laboratory conditions, maze-bright and maze-dull rats showed significant differences in the number of errors they made when learning to navigate through a maze. However, when both types of rats were raised in environmentally impoverished conditions, maze-bright rats appeared to lose their advantage, making the same number of errors as maze-dull rats. Conversely, when both types of rats were raised in an enriched environment, maze-dull rats gained a significant advantage, reducing their number of errors to a level almost equal to that in maze-bright rats. (Data from Cooper and Zubek, 1958.)

heritance patterns provided substantial evidence for the theory that general intelligence is influenced by numerous genes.

More recently, one group of researchers produced transgenic lines of mice by altering a gene that controls production of **N-methyl-D-aspartate** (**NMDA**) receptor proteins (Tang et al., 1999). NMDA receptors bind the amino acid **glutamate**, which acts as an excitatory neurotransmitter, and they have been implicated in long-term memory consolidation and storage. Mice in one transgenic line created by Tang and colleagues overexpressed NMDA receptors in the forebrain region and performed significantly better than wild-type mice in several cognitive tasks, including the Morris water maze (Figure 9.6, Box 9.1). This study simultaneously suggested a specific candidate gene and a neurological basis for this feature of intelligence. In addition, it demonstrated the feasibility of manipulating an organism to enhance a genetic trait within the normal range of function. This latter achievement has profound implications when human applications are considered. More consideration is given to the potential for enhancing human behaviors within the normal range of function in the final chapter of this text.

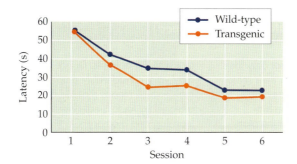

FIGURE 9.6 Latency (time taken to locate the hidden platform) in the Morris water maze for wild-type and transgenic mice. The transgenic mice had a gene alteration that enhanced NMDA receptor production, and thus had a greater forebrain concentration of NMDA receptors. Wild-type mice took significantly longer to find the platform than did transgenic mice in sessions 3 and 4. These results show more rapid maze learning and more efficient use of spatial cues by the transgenic animals. (After Tang et al., 1999.)

BOX 9.1 The Morris Water Maze

Chapter 6 described several experimental devices used for testing fear in rodents. Just as researchers use a variety of devices to measure fear and anxiety in rodent models, they use a variety of mazes and operant tasks to quantify intelligence. As noted earlier, intelligence is a rather abstract concept, but it is generally considered to be measurable using tests of cognitive abilities, such as visual and spatial memory. Among the most popular tests of these two cognitive abilities is the **Morris water maze**. Developed by Richard G. Morris at the University of Edinburgh in 1984, the test requires a rat or mouse to find a hidden platform submerged just below the surface in a large vat of water (Figure). The researcher places the animal in the water, which has been made cloudy (powdered milk is sometimes used) so that the animal cannot see the platform. Over repeated trials, the animal learns to use spatial cues surrounding the vat as landmarks to guide its movements to the platform. By learning and remembering

these visual cues, the animal is eventually able to swim directly to the platform, regardless of where it is initially placed in the vat. Interestingly, if researchers are present in the area of the vat during testing, they need to remember to stand in the same place for each trial, as they themselves will become landmarks used by the animals.

Data from the Morris water maze may be analyzed in a variety of ways. Latency from the time of placement in the vat to the time of finding the platform is a primary measure of learning. The platform may also be removed after initial learning trials. With this manipulation, researchers can compare the amount of time the animal spends swimming in the quadrant where the platform was located with the amount of time it spends in the other three quadrants. Animal placement location or orientation at the beginning of the task can be consistent or varied between trials. With consistent placement

(Continued on next page)

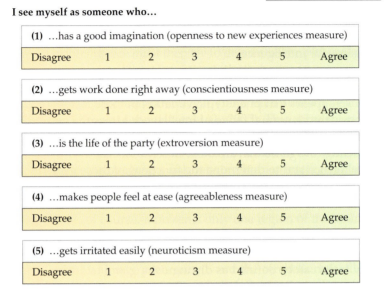

I see myself as someone who…

(1) …has a good imagination (openness to new experiences measure)

| Disagree | 1 | 2 | 3 | 4 | 5 | Agree |

(2) …gets work done right away (conscientiousness measure)

| Disagree | 1 | 2 | 3 | 4 | 5 | Agree |

(3) …is the life of the party (extroversion measure)

| Disagree | 1 | 2 | 3 | 4 | 5 | Agree |

(4) …makes people feel at ease (agreeableness measure)

| Disagree | 1 | 2 | 3 | 4 | 5 | Agree |

(5) …gets irritated easily (neuroticism measure)

| Disagree | 1 | 2 | 3 | 4 | 5 | Agree |

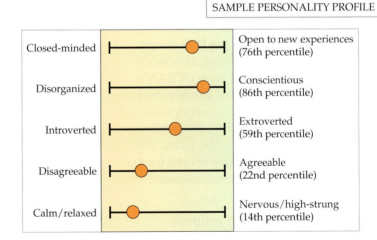

SAMPLE PERSONALITY PROFILE

Closed-minded	Open to new experiences (76th percentile)
Disorganized	Conscientious (86th percentile)
Introverted	Extroverted (59th percentile)
Disagreeable	Agreeable (22nd percentile)
Calm/relaxed	Nervous/high-strung (14th percentile)

FIGURE 9.9 Examples of questions and an individual personality profile from the FFM rating scale used to assess the Big Five personality traits. Scoring of answers to 50 such questions yields percentile rankings within a continuum of scores representing the general population. To evaluate your own personality traits, visit www.outofservice.com/bigfive/.

tors that shape emotional and cognitive expression. The influence of environmental experiences on temperament leads to predictable patterns of behavioral expression. The term **personality** is used to describe these biologically based, environmentally shaped, traits. In fact, temperament, rather than personality, is frequently the preferred term when studying children, who express many behaviors that are not significantly modified by environmental factors. The clear relationship between these two terms explains their analogous use in studies comparing humans and animals.

Just as human personality can be defined as the sum of several core traits, animal temperament can be evaluated by assessing a variety of related behavioral traits. In dogs, for example, Hart and colleagues (Hart and Hart, 1985; Hart and Miller, 1985) identified 13 distinct behaviors that could be quantified to generate a temperament profile that could be matched with pet owners' needs and expectations (Table 9.1). Some of these 13 behaviors probably represent canine variations of the OCEAN traits evaluated in the FFM. Further, just as human traits may be quantified to create a personality profile, Hart et al. created temperament profiles for over 50 breeds of dogs (Table 9.2). The consistent expression of temperament traits in these inbred animals roughly parallels the shared personality traits seen in humans, particularly first- and second-degree relatives.

More recently, van Oers et al. (2004) devised a series of tests to assess boldness and risk-taking behavior in the great tit, a small songbird (Figure 9.10). These researchers then used trait breeding techniques to create a line of low-risk birds and a separate line of high-risk birds. Their results showed significant differences in risk-taking behavior between the two lines that were apparent in the first generation of trait breeding, and increasing differences were noted in subsequent generations. The premise and design of this study may seem familiar, as it closely resembles the seminal work on maze-bright and maze-dull rats. In the van Oers et al. study, however, the findings show heritability for a temperament trait, rather than a cognitive ability. The traits they measured in birds could be homologous to human personality characteristics associated with extroversion, openness to new experiences, or neuroticism.

TABLE 9.1	Behavioral Traits Used to Rate Temperament in Dogs
TRAITS	
Excitability	
General activity	
Snapping at children	
Expressive barking	
Affection demand	
Territorial defense	
Watchdog barking	
Aggression to dogs	
Dominance over owner	
Obedience training	
Housebreaking ease	
Destructiveness	
Playfulness	

Source: Hart and Hart, 1985; Hart and Miller, 1985.

BOX 9.2   The Price of Promiscuity

Although female sexual promiscuity is frowned upon in many human cultures, it might seem that for other animals, promiscuous behavior would offer an evolutionary advantage. In at least some species, that does not appear to be the case.

In many animals, fertilization is enhanced by the presence of a seminal plug that is formed in the vaginal cavity immediately after a male ejaculates. Dislodging this plug too soon after insemination reduces the chance of fertilization (Sachs, 1982). One means of dislodging a seminal plug is by engaging in copulation. Think about the implications of this process in light of what you now know about natural selection. If a female rat copulates with several males in rapid succession, each successive mating partner will dislodge the seminal plug produced by the previous partner, reducing the likelihood that the previous partner's sperm will fertilize the female's eggs. This means that the last mating partner is as likely as the first (or perhaps more

likely) to father the female's offspring. Now think about which male is likely to be the last in line to copulate. If you guessed the oldest, most sickly, weakest, or otherwise most inferior male, you are correct. This would mean that the least fit male would have the potential to pass along his genes to the promiscuous female's offspring. The survival of those offspring might be hindered by his inferior genes, reducing the female's reproductive success.

Taken together, this information suggests that hypersexual behavior in a female rat would decrease her chances of passing on her genes to future generations. Therefore, it may not surprise you to learn that there is a low occurrence of hypersexuality in this species, forming one of the tails of the normal distribution of sexual behavior. The other tail of the distribution is much more intuitive. At the low end of the sexual activity scale is asexual behavior, which reduces the likelihood of gene transmission for obvious reasons.

Influence of genes on homosexual behavior

In terms of behavior genetics, homosexuality is a highly intriguing trait. Once widely considered a pathological condition, it was described by the *Diagnostic and Statistical Manual of Mental Disorders* as a treatable mental illness until the 1970s. It was not until 1973 that the American Psychiatric Association voted to remove homosexuality from its list of mental disorders. Even then, an alternative disorder title of **Ego Dystonic Homosexuality** (**EDH**) was included in *DSM-III*, published in 1980. EDH was defined as a disorder in which intrusive homosexual urges produce anxiety or dysphoria (the opposite of euphoria) or otherwise impede normal levels of expression of heterosexual behavior. However, even this classification was eliminated from the revised version of *DSM-III* (*DSM-IIIR*) in 1986. Today, most researchers do not view homosexual behavior as an abnormal expression of heterosexual traits. Instead, homosexual traits are generally regarded as independent of, but related to, heterosexual traits. This approach is supported by the observation that many individuals display both heterosexual and homosexual tendencies without exhibiting the level of distress that would be expected to occur with abnormal expression of a single trait.

Several factors point to a polygenic basis for homosexual behavior. The most apparent factor is the lack of qualitative expression. In other words, sexual behavior is not expressed as exclusively homosexual or exclusively heterosexual. Between these two exclusive end points, there is a wide range of bisexual behavior, ranging from very limited to very extensive (but not exclusive) homosexual trait expression. A second feature pointing to polygenic mediation of homosexual behavior is a lack of Mendelian patterns of inheritance. This point, however, needs some further explanation. A person whose sexual behavior was exclusively homosexual would not, by definition, produce offspring. Such a limitation, of course, would greatly reduce the number of possible study subjects expressing that trait in successive generations. However, the possibility that homosexuality is a single-gene trait can be eliminated by assessing monozygotic twins, who do not show a 100% concordance rate for this phenotype.

Although they have ruled out the possibility that trait expression being controlled by a single gene, twin studies do show a significant level of heritability for homosexuality, particularly in males. Among the most stringently controlled of these studies is that of Whitam and colleagues (1993), who reported a 65.8% concordance rate for homosexuality in monozygotic twin pairs compared with a 30.4% concordance rate for dizygotic twins. Interestingly, very little influence of shared environment on the expression of homosexual behavior has been found in any twin study. There is, in fact, no evidence to support the idea that being raised by homosexual adoptive parents has any influence on the development of this trait in adopted children.

Just like heterosexual behavior, homosexual behavior is normally distributed across the range of phenotypic expression. In other words, varying levels of sexual activity are seen for both heterosexual and homosexual individuals, with the greatest number reporting a moderate range of expression. This finding further supports the theory of multiple gene influence on the expression of homosexual behavior.

Although psychologists now almost universally accept homosexual behavior as a variation in sexual expression that is not considered pathological, why such a trait should exist is an obvious question. After all, on the surface, it would appear that homosexual behavior is counterproductive to reproductive success, especially considering that the time and energy spent in expression of this trait becomes unavailable for reproductive activities. However, there are several theories that could explain this apparent paradox.

It is possible that homosexual behavior is influenced by genes that serve other basic and vital functions. In this case, if the same genes that contribute homosexual behavior also in some way enhance reproductive success, they could be sustained from one generation to the next. Camperio-Ciani and colleagues (2004) recently used pedigree studies to show that maternal lines in the families of homosexual men produced significantly more offspring than did maternal lines in the families of heterosexual men. Think about this finding for a moment. If a particular allele promotes male homosexual behavior, the theory of natural selection would predict its elimination over time. However, perpetuation of that same allele would be expected if it also increased childbearing success in women.

It is also possible that some genes promote attraction to males of a species, while other genes promote attraction to females of a species. The function of such genes might normally be influenced by an interaction with reproductive hormones (i.e., expression of "male attraction" genes is promoted by estrogens, whereas expression of "female attraction" genes is promoted by testosterone). In this case, an altered gene that did not respond, or responded inappropriately, to hormones could result in homosexual behavior patterns.

There is also an increasingly popular theory that, though they do not directly enhance reproductive success, homosexual behaviors may still be important for the success of some species. This theory has been formulated largely from observations of animal models. **Intragender sexual action**, a term sometimes used to describe homosexual behavior in animals, has been observed in a large number of species. Interestingly, this trait is not considered detrimental to reproductive success, and some researchers have argued that it could actually contribute to the overall success of a species (see Sommer and Vasey, 2006). For example, animals may use intragender sexual action to establish dominance hierarchies, particularly in highly social species. This activity, as opposed to violent physical fighting, reduces the risk of injury or death for competing males, benefiting the group as a whole. Intragender sexual action is also thought to enhance both male and female bonding in some species, resulting in greater cooperation and efficiency within the group.

The idea that some level of homosexual behavior could be beneficial to a group of animals is somewhat controversial, but it is difficult to find any other explanation for the trait's existence in such a vast array of animals that have evolved so successfully. Although the benefit of this trait in humans may not be obvious, the implication of genetic influence suggests that homosexual behaviors could have played a role in the successful evolution of our own species. In this regard, researchers should find it beneficial to view homosexuality as they would other traits that are expressed within a normal range. Removing this trait from the DSM listing of mental disorders in the 1970s represented an important step in advancing the study of homosexual behavior. Removing the label of "pathology" from this trait has prompted many behavior genetics researchers to refocus their studies, shifting their search for abnormalities in heterosexual gene function to a more viable search for candidate genes contributing to a separate but related behavioral trait.

Chapter Summary

Rather than using the terms "normal" and "abnormal," most researchers prefer to regard behavioral expression as being within or outside of a normal range. Assessment of cognitive ability with IQ scores is a good example of the use of a standardized quantitative rating scale to define such a normal range. Because quantitative traits are normally distributed across the range of phenotypic expression, researchers can use a statistically defined distribution to quantify the normal range. Personality traits and mood may also be quantified using scales such as the Behavior Inhibition Scale (BIS) and the DEpression and EXhaustion (DEEX) scale, respectively.

The terms "intelligence" and "cognitive abilities" are closely intertwined and probably refer to features of a common trait. The most popular explanation of the relationship between intelligence and cognitive abilities is based on the psychometric theory, which proposes that intelligence is a combination of abilities that can be measured by testing multiple cognitive functions. Charles Edward Spearman, an early advocate of psychometric theory, coined the term "general intelligence factor" (g) to describe a common element that contributes to all tests of intelligence. The relationships of individual cognitive abilities to g, however, are not equal. In fact, the most accurate assessments of intelligence are believed to be obtained using factorial analyses, which give greater weight to some cognitive abilities than to others.

Among the first systematic experiments on breeding for intelligence in animals were studies of maze-bright and maze-dull rodents conducted beginning in the 1930s. Although those early studies showed that intelligence has a strong genetic component, subsequent studies found that environmental impoverishment and enrichment could significantly affect cognitive abilities regardless of genetic predisposition. Most recently, researchers have created transgenic lines of mice with superior maze-learning ability by altering a gene that controls expression of NMDA receptor proteins.

Studies of human intelligence have shown correlations in test scores ranging from 0.66 to 0.88 for monozygotic twins and from 0.42 to 0.62 for dizygotic twins, indicating a significant genetic influence on intelligence. Research on human brains has shown that frontal lobe volume is positively correlated with cognitive abilities. This brain region contains a relatively high concentration of NMDA receptors, a finding that nicely parallels the relationship between NMDA receptor gene expression and cognitive abilities found in rodents.

Personality traits are distinguishing and enduring qualities or characteristics of an individual's behavior. The predominant personality trait theory in behavior genetics research is the Five Factor Model (FFM). A core feature of the FFM is that it does not identify individuals as possessing specific personality types. Instead, it rates them on a continuum of possible scores for each of five traits to produce a personality profile. Animal models are also used in the study of personality traits, though the term "temperament" is typically used in place of "personality." In dogs, 13 distinct behaviors, some of which resemble human personality traits evaluated in the FFM, have been quantified to generate a temperament profile, and over 50 breeds of dogs have been profiled. Other animals, including birds, have also been used to assess the heritability of personality traits. Modern molecular techniques are now being used to link human personality traits to animal temperament traits. For example, an allele of the *D4DR* gene that is associated with high novelty seeking scores in humans closely resembles a *D4DR* allele found in one breed of playful and highly trainable dogs.

Researchers have employed twin and adoption studies to assess genetic influences on personality traits. Results from self-rating scales reveal that genetic influences account for approximately 30%–50% of the observed variation in several different personality traits. One recent study concluded that nearly all personality traits were influenced approximately 25% by shared environment, 35% by unshared environment, and 40% by genetics.

Sexual behavior appears to be a complex trait regulated by numerous genes. Several brain structures and neurotransmitters that directly mediate sexual behavior have been identified. In rodents, for example, dopamine activity in the medial preoptic area (MPOA) regulates aspects of male sexual drive. In human males, the *DRD4* dopamine receptor gene has been similarly linked to sexual arousal, desire, and function. In one study of twins, concordance rates for reports of multiple sex partners were 36.9% for monozygotic twin pairs and 26.1% for dizygotic twins, suggesting a genetic influence on this realm of sexual behavior.

Homosexual behavior is typically studied as independent of, but related to, heterosexual behavior. Several factors point to a polygenic basis for homosexual behavior, including a lack of qualitative expression, lack of simple Mendelian patterns of inheritance, and a monozygotic twin concordance rate well below 100%. However, twin studies do suggest that homosexual behavior has a significant genetic component, with concordance rates of 65.8% in monozygotic twin pairs compared with 30.4% in dizygotic twins. Shared environment appears to have little effect on the expression of homosexual behavior, quelling arguments that being raised by homosexual adoptive parents may influence development of this trait in adopted children. One theory proposed to explain the persistence of homosexual behavior throughout evolution is that certain species (including humans) may benefit from expression of some homosexual traits. For example, animals may use intragender sexual action as an alternative to violent physical encounters as a means of establishing dominance hierarchies. Intragender sexual action is also thought to enhance both male and female bonding in some species, resulting in greater cooperation and efficiency within the group.

Review Questions and Exercises

- Define cognitive abilities and intelligence, and explain the relationship of *g* to both.

- Researchers demonstrated long ago that the cognitive abilities involved in maze learning were heritable traits. Explain how environmental factors can influence these particular traits.

- Animal research has revealed a link between genes that regulate NMDA receptors and cognitive abilities. Describe the recent findings from human research that appear to parallel this animal work.

- List the five personality traits assessed using the Five Factor Model (FFM). How is the FFM used in assessing an individual's personality?

- Compare and contrast the methods and results of the classic studies of maze-bright and maze-dull rats with those of more recent studies of boldness and risk-taking in birds.

- What evidence exists to suggest that features of male sexual behavior are heritable? What neurotransmitter system and corresponding gene have been implicated in both animal and human studies on this topic?

- Explain how the approach to researching homosexual behavior has changed over the past several decades. Describe the current perspective taken by most behavior genetics researchers when studying homosexual behavior.

- From an evolutionary perspective, it could be argued that homosexual behavior expends time and energy that could otherwise be used to increase reproductive success. Explain a potential benefit of homosexual behavior that may explain why this behavior persists in some species.

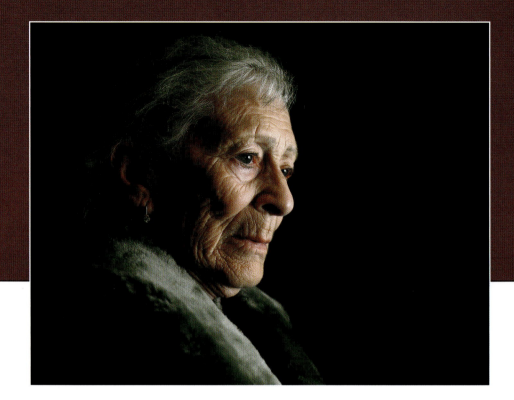

10

Primary Cognitive Disorders

Chapter 9 gave a brief summary of evidence showing that intelligence is influenced by genetic factors. One conclusion that is clear from both human and animal studies on the normal range of intelligence is that this trait is influenced by numerous genes regulating many different cognitive functions. In addition, genes are believed to influence a more general intellectual trait that affects all cognitive skills to varying degrees. Once it is understood that intelligence is a complex trait that encompasses a variety of cognitive abilities, each controlled by numerous genes, it becomes easy to imagine a wide range of potential genetic aberrations that could cause intellectual deficits. When these deficits are mild, the result is an individual whose intelligence measures are in the low normal range. As these deficits become more pronounced, however, individuals may be diagnosed with mental retardation, dementia, or learning disabilities. Each of these diagnoses indicates some disorder marked primarily by cognitive dysfunction. This chapter emphasizes the possible genetic influences on and heritability of these abnormal phenotypes. One common thread connecting all three categories is the assumption that a greater knowledge of genetic influences on the normal range of cognitive function will enhance our understanding of these areas of cognitive impairment.

FIGURE 10.1 Tom Hanks won an Academy Award for Best Actor in the 1994 movie *Forrest Gump*. In the movie, Hanks portrayed a thoughtful, ambitious, and sometimes creative individual with borderline intelligence. Though purely fictional, the character of Forrest Gump demonstrates how an individual with low cognitive functioning might fail to satisfy the criteria for mental retardation.

Mental Retardation

Mental retardation is described as a condition of low general intellectual function that includes impaired abilities to acquire skills required for normal daily activities. These characteristics must be apparent before the age of 18 for a diagnosis of mental retardation, as they closely parallel the symptoms of dementia seen in numerous adult-onset degenerative disorders (discussed in the next section of this chapter). One problem with this definition is the term "low general intellectual function," which itself is ill defined. In Chapter 9, it was noted that the normal range of human intelligence, as measured by IQ scores, is frequently cited as 85–115. An IQ score below 70 is generally used as the threshold for a diagnosis of mental retardation. This leaves a rather wide range of scores that are not categorized as being either within the normal range or within the range of mental retardation. IQ scores in the range of 70–85 are often referred to as **borderline**, indicative of the "gray area" seen when attempting to assign a qualitative diagnosis to a quantitative phenotype influenced by numerous genes. Understandably, among those scoring within this borderline range, individual variation makes some appear obvious candidates for a mental retardation diagnosis, whereas others may not clearly meet the diagnostic criteria given above (Figure 10.1). Thus, a certain amount of leeway is allowed for clinical diagnosis of those whose test scores fall in the borderline range.

Severity range

Generally, a person who scores below 70 on an IQ test exhibits cognitive impairment severe enough to meet the criteria for a diagnosis of mental retardation. In addition, it is evident that IQ scores within the range of mental retardation are directly correlated with levels of cognitive functioning. The strength of this correlation, along with the range of the phenotype, makes it practical to subdivide mental retardation into severity categories for diagnostic and clinical applications.

- **Mild mental retardation** (IQ = 55–69) is characterized by the ability to learn many practical skills, including reading, writing, and arithmetic, at a grade-school level. Individuals scoring in this range often show reasonable degrees of social and job skills and are capable of living independently with some guidance and supervision.

- **Moderate mental retardation** (IQ = 40–54) is characterized by noticeable motor skill impairment and speech deficits. Reading, writing, and arithmetic skills are very limited or absent. Individuals scoring in this range generally undertake only simple tasks accomplished with close supervision or in a highly controlled environment. They also typically live with a caretaker.

- **Severe mental retardation** (IQ = 25–39) is characterized by limited communication and motor skills. Individuals scoring in this range require training for acquisition of basic hygiene and feeding skills. As adults, they are usually confined to a protective environment marked by close supervision and consistent daily routines.

- **Profound mental retardation** (IQ < 25) is frequently comorbid with other medical problems and is characterized by severe communication and motor impairment. Individuals scoring in this range require assistance in nearly all aspects of daily functioning, including feeding and personal hygiene.

Prevalence

Mental retardation may be caused by a variety of factors, including genetics, problems in fetal development, problems during birth, and problems in early childhood development (Table 10.1). **Teratogen** is a general term that is sometimes used to refer to any chemical, drug, or other substance (e.g., heavy metals, viruses) known to produce mental retardation or other birth defects when present during fetal development. Some teratogens may induce developmental deficits by themselves, whereas others may produce effects only in genetically predisposed individuals. The potential for such gene–environment interaction makes it difficult, in some cases, to judge the relative influence of genetics on mental retardation and other birth defects.

The prevalence of mental retardation is between 1% and 3% in the general population, with the highest estimates based on IQ scores only and lower estimates based on stricter diagnostic criteria. Although the incidence is typically higher in developing countries, this higher rate is not a result of genetic differences. Rather, in developing countries or economically challenged regions of industrialized countries, environmental factors associated with poverty and inadequate health services increase the likelihood of problems associated with fetal development, birth, and childhood development.

For the purpose of this text, mental retardation resulting from genetic causes is of greatest interest. However, establishing the prevalence of genetically based mental retardation is not easy. For example, in an assessment of 151 individu-

Rett syndrome is caused by mutation of a single gene, located on the long arm of the X chromosome (Xq28), that produces **methyl-CpG-binding protein 2 (MeCP2)** (see Bienvenu et al., 2000). MeCP2 is a regulatory protein that directs the production of several other proteins. As you might guess, disruption of a protein with such global function has profound effects on development throughout the entire body, including the central nervous system. Males who inherit a mutant allele of the *MeCP2* **gene** are typically aborted during fetal development, making Rett syndrome almost exclusive to females. In females who inherit this same allele on one of their two X chromosomes, production of MeCP2 is thought to be taken over, although in incomplete fashion, by the unaffected X chromosome. In those rare cases in which a female has two affected alleles (one on each X chromosome), the result is similar to that in a male who inherits one allele: spontaneous abortion occurring during fetal development. In females who inherit a single Rett syndrome allele, symptoms generally do not arise until the child reaches 6 to 18 months of age, and may include moderate to severe mental retardation, poor muscle control, severe language impairment, loss of social engagement, and respiratory difficulties.

Phenylketonuria (PKU), like Rett syndrome, is a single-gene disorder that results in moderate to severe mental retardation. Its adverse effects, however, may be prevented by a strict dietary regimen that excludes phenylalanine, as described in Chapter 5. PKU is now considered a treatable disorder, but before the discovery of the dietary treatment, it was among the leading causes of mental retardation.

As the name implies, **microdeletion syndromes** are caused by losses of DNA segments from a chromosome during recombination. The result is a chromosome with enough genetic material to sustain development through birth, but enough missing genetic material to produce observable deficits. Over 80% of cases of **DiGeorge syndrome** are attributed to a deletion from the long arm of chromosome 22 (22q11) (Driscoll et al., 1993). A small number of cases have been attributed to microdeletions from other chromosomes (most notably chromosomes 10 and 18). In cases of 22q11 deletion, developmental delays and learning disabilities, particularly difficulties with speech and language acquisition, are common, and nearly half of all patients are diagnosed with mental retardation. Common physical abnormalities include cardiac malformations, cleft palate, and decreased muscle tone.

Prader-Willi syndrome and Angelman syndrome are two disorders with distinctly different characteristics, including different severities of mental retardation, that are caused by a microdeletion of the same region of chromosome 15 (15q11–q13). These disorders were described in Chapter 5 as examples of genomic imprinting, in which phenotypic expression depends on which parent contributes the mutant allele. Inheritance of this microdeletion from the father results in Prader-Willi syndrome, but when this same microdeletion is inherited from the mother, the result is Angelman syndrome.

Table 10.3 summarizes some of the information related to the disorders associated with mental retardation listed in this chapter. Advances in the Human Genome Project, in conjunction with improved methods for visualizing DNA, will undoubtedly allow researchers to identify additional causes of mental retardation in the near future. As noted earlier, many cases of mental retardation are currently categorized as idiopathic. Progress in the area of behavior genetics will surely move many cases of idiopathic mental retardation into the genetic category, further demonstrating the importance of understanding gene and chromosome influences on cognitive development.

TABLE 10.3 Chromosome and Gene Abnormalities Underlying Mental Retardation

CATEGORY	EXAMPLE	CAUSE/ LOCATION	ASSOCIATED PROTEIN	HERITABILITY	NOTES
Trisomy	Down syndrome	Nondisjunction (chromosome 21)	Many	Very limited	Incidence of trisomy is correlated with age of mother at conception
	Edward syndrome	Nondisjunction (chromosome 18)	Many	None	
Triplet repeat expansion	Fragile X syndrome	Addition of genetic material (Xq27.3)	FMR protein	Affects 2x more males than females	Premutations seen in some unaffected parents
Single-gene mutations	Rett syndrome	MeCP2 (Xq28)	Methyl-CpG-binding protein	Usually lethal in males	2nd leading cause of mental retardation in females
	PKU	PKU (12q24.1)	Phenylalanine hydroxylase	Single recessive allele	Treatable with early dietary restriction
Microdeletion	DiGeorge syndrome	Deletion at 22q11	Unclear: Possibly G protein	Mendelian pattern found in rare cases	Deletions from 10th and 18th chromosomes underlie rare cases
	Prader-Willi and Angelman syndromes	Deletion at 15q-11–15	Unclear: Probably include Necdin and Magel2	None	Disorders represent example of genomic imprinting

Heritability

Earlier in this chapter, it was noted that Airaksinen and colleagues (2000) identified a genetic basis in four times as many cases of severe mental retardation as in cases of mild mental retardation. On the surface, these data seem to suggest that genetic aberrations are associated with the most pronounced forms of retardation, whereas other factors (perhaps environmental) produce more subtle forms. However, this is probably not the case. Although it is true that severe mental retardation is frequently traced to a genetic cause, it is also true that current research methods reliably detect only the most prominent genetic abnormalities, such as trisomy and triplet repeat expansion. In cases in which less pronounced genetic changes have been identified, such as single-gene mutations or microdeletions, the prevalence of the disorders has been great enough to allow successful screening of a large group of affected individuals.

One theory proposed to explain the large number of cases of mild mental retardation that retain the idiopathic label is that they merely reflect the small percentage of low-IQ individuals expected within the normal distribution of intelligence. More specifically, if normal intellectual development arises from the influence of many genes working together (referred to in Chapter 6 as quantitative trait loci, or QTL), then some small percentage of individuals would be expected to inherit a combination of genes that would result in IQ scores below 70 (just as a small percentage score above 130). Although this theory seems reasonable, Spinath and colleagues (2004) recently conducted an extensive review of twin data, from which they concluded that the prevalence of mild mental retardation exceeds what would be expected in the normal distribution of intelligence. Part of the explanation for this finding comes from the fact that mild mental impairment is a feature shared by a wide range of rare single-gene disorders. Although such disorders increase the frequency of low IQ scores, they do not contribute equally to the entirety of the distribution curve (i.e., no known single-gene disorder is characterized by IQ scores that are significantly above average).

As noted in the previous section, several single-gene disorders characterized by mental retardation are inherited on the X chromosome. In some of these disorders, such as fragile X syndrome, the disorder is expressed predominantly in males, because females possess an unaffected gene on the homologous chromosome that is usually capable of compensating for the effects of the abnormal allele. In other X-linked disorders, such as Rett syndrome, the prevalence is greater in females than in males. A greater incidence of a disorder in females often suggests the genetic defect produces highly dysfunctional effects severe enough to have lethal consequences during male fetal development. Females, however, may survive to birth by partially compensating for the defective allele with a matching, normally functioning, gene on the second X chromosome. This compensation increases the number of females compared with males who are born with the defect and thus with mental retardation.

Microdeletions typically occur during recombination, as described in Chapter 5. When segments of DNA literally break off from one chromosome and fuse to the homologous chromosome, it is possible for small segments to be lost and subsequently excluded from the recombined chromosome. Although such a process in itself is not heritable, it does appear that once integrated into the genome, some microdeletions begin to follow a Mendelian pattern of inheritance. For example, several cases of **familial DiGeorge syndrome** have been reported (Leana-Cox et al., 1996).

In addition to familial DiGeorge syndrome, several genetic disorders that follow a normal Mendelian pattern of inheritance are characterized by mental retardation. Two unaffected carriers of the mutated PKU allele, for example, have a 50% chance of producing offspring who are also unaffected carriers, and a 25% chance of producing a child who exhibits the phenotype. **Neurofibromatosis type 1 (NF1)** is a dominantly inherited single-gene disorder that affects about 1 in 3,000 individuals in the United States (Friedman, 1999). NF1 is characterized by multiple café-au-lait spots (dark pigmented patches) and neurofibromas (benign tumors created by abnormal growth of glial cells) on or under the skin (Figure 10.5). In addition to these peripheral features, mental retardation is diagnosed in individuals with NF1 at a rate slightly higher than in the general population. One-half of the offspring of an individual affected by NF1 are expected to develop the disorder. The *NF1* gene is located on the long arm of chromosome 17 (17q11.2).

Clearly, the range of possible defects that can produce mental retardation makes analysis of hereditary patterns a difficult task. Although the examples given here show the basis for some reliable patterns, advances in technology will surely increase our understanding of the heritable nature of more forms of mental retardation in the future.

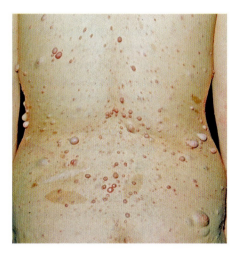

FIGURE 10.5 Neurofibromatosis type 1 (NF1) is characterized by neurofibromas (benign tumors that develop from glial and connective tissue) on or under the skin. In addition, pigmented regions called café-au-lait spots (literally, "coffee with cream," which accurately describes their color) are apparent. About half of all NF1 patients exhibit learning disabilities. In some cases, mental retardation is also diagnosed.

A third proposal is that multiple genes contribute to a single cognitive process that influences all measures used to diagnose dyslexia (Figure 10.9C). For example, ability to match letters to appropriate sounds probably influences all measures of reading ability. Gene combinations that differentially contribute to this particular cognitive process could then produce the continuum of deficits seen in dyslexia. This proposal eliminates the oversimplified view of single-gene control of cognitive processes and the assumption that individual test scores represent independent cognitive abilities. The most apparent problem with this model is the assumption of a single cognitive process influencing all the test measures. The fourth and final proposal is that multiple genes influence multiple cognitive processes, each of which has some degree of influence on multiple measures of reading ability (Figure 10.9D). This model is the most complex of the four, and probably the most accurate, considering the likely influence of genes on multiple brain regions and the overlap of individual cognitive abilities when applied to more general reading tasks.

Figure 10.9 shows part of the difficulty in establishing a definitive hereditary basis for learning disabilities. As noted by Fisher and DeFries, even the most complex scenario is a "gross oversimplification" because it does not consider either known environmental effects or the likely interactions between the genes involved.

In spite of the obvious difficulty of interpreting the genetic basis of learning disabilities, there is little question that dyslexia, like many other learning disabilities, is heritable. More specifically, children of a dyslexic parent are at greater risk of developing this learning disability than are children of nondyslexic parents (Wolff and Melngailis, 1994). Furthermore, both the risk of developing dyslexia and the severity of the impairment were found to be even greater in children of two dyslexic parents. Such reports of high parent–offspring concordance rates are highly suggestive of a genetic influence, but recall that such comparisons do not exclude the possibility of environmental effects. For example, parents with impaired reading skills could, in theory, "teach" such deficits to their children by example. This possibility has been effectively refuted, however, in at least one extensive twin study that showed significantly greater concordance rates between monozygotic (68%) twin pairs than between dizygotic (38%) twins (DeFries and Alarcon, 1996). These data not only indicate a lack of shared environmental effects, but also suggest that dyslexia is influenced by numerous genes, rather than by a single gene, as indicated by the lack of a concordance rate nearing 100% for monozygotic twins.

There is at least one report of a form of dyslexia that has followed a dominant Mendelian inheritance pattern. Nopola-Hemmi and colleagues (2001) identified the associated allele on chromosome 3, where it was localized to the same region as the *ROBO1* gene. This dominantly inherited form of dyslexia appears to be caused by a unique allele that includes the *ROBO1* gene, directly affects the *ROBO1* gene, or affects the same regulatory processes controlled by the *ROBO1* gene.

Additional research will be needed to better understand the genetic basis for dyslexia and learning disabilities in general. Difficulties in establishing clearly defined parameters for diagnosing these disorders have created related problems in identifying specific cognitive processes that underlie the deficits. These problems, in turn, have made it difficult for researchers to seek out candidate genes. Heritability studies suggest that several genes influence dyslexia, with the most likely candidate genes affecting neural development. It is likely that, in time, other less prevalent learning disabilities will generate similar findings. With advances in the technologies used to delineate the human genome, more data are expected from this area of research in the near future.

Mental retardation, dementia, and learning disabilities represent three categories of primary cognitive dysfunction. Phenotypic expression in these three categories ranges from profound global retardation of cognitive function, resulting in very low IQ scores, to highly specific deficits in reading comprehension, which can occur in individuals with genius-level IQ scores. The time of onset of these phenotypes has a similarly wide range, from prenatal development to late adulthood. It is hoped that future research in the areas of both normal and abnormal cognitive development will contribute to a common body of data that can be used to delineate genetic influences on these broad categories of cognitive disorders.

Chapter Summary

Mental retardation is a condition of low general intellectual function typically characterized by an IQ score below 70. Some leeway is allowed in the clinical diagnosis of individuals with scores in the borderline range of 70–85, between the normal range and mental retardation. Within the range of mental retardation, there are four categories of severity based on degree of impairment. Mild mental retardation (IQ = 55–69) is characterized by the ability to learn practical skills at a grade-school level; moderate mental retardation (IQ = 40–54) by reading, writing, and arithmetic skills that are very limited or absent; severe mental retardation (IQ = 25–39) by limited communication and motor skills; and profound mental retardation (IQ < 25) by severe communication and motor impairment. The prevalence of mental retardation is between 1% and 3% of the general population. Its prevalence is higher in developing countries where environmental factors are more likely to negatively affect fetal development, birth, and childhood development.

The prevalence of genetically based mental retardation is difficult to estimate, though one recent study found that 28% of cases in a large sample were caused by chromosomal or genetic aberrations. When that sample was subdivided by severity, genetic causes were identified in 44% of severe cases compared with 11% of mild cases. It is also worth noting that these percentages are probably underestimations, since undiscovered genetic factors may contribute to many cases now listed as "cause unknown."

Down syndrome (also known as trisomy 21), which produces mild to moderate mental retardation, is caused by the addition of a third chromosome 21 to the genome. Trisomy results from chromosomal nondisjunction during meiosis. The likelihood of nondisjunction is directly and positively correlated with maternal age, explaining why the prevalence of Down syndrome births increases significantly among women bearing children in the fourth and fifth decades of life. Other trisomy disorders include Edward syndrome and Patau syndrome, which both result in profound mental retardation in those rare cases in which the affected individual survives beyond the first year of life.

Other forms of mental retardation caused by genetic aberrations include fragile X syndrome, caused by expansion of a CGG repeat on the X chromosome. Premutation is a phenomenon sometimes seen in such expanded triplet repeat disorders in which an unaffected parent possesses a lesser variant of the expansion that affects the offspring. Rett syndrome, characterized by moderate to severe mental retardation, is also caused by a mutation on the X chromosome. Rett syndrome is found almost exclusively in females, as the presence of this mutation in males is lethal. Microdeletion syndromes, caused by losses of DNA segments during recombination, include DiGeorge syndrome, Prader-Willi syndrome, and Angelman syndrome. These latter two syndromes are caused by a microdeletion from the same region of chromosome 15 that, when inherited from the mother, results in Angelman syndrome and when inherited from the father, results in Prader-Willi syndrome.

The prevalence of dementia increases with age. About 1% of people 60–65 years of age exhibit dementia, and the number of affected individuals then doubles with each successive 5-year interval. The most common form of dementia is associated with Alzheimer disease, accounting for more than half of all diagnosed cases. Other forms of dementia include vascular dementia, usually resulting from stroke (or a series of small strokes), Parkinson disease dementia, and Lewy body dementia, which some researchers believe may be comorbid expression of Parkinson and Alzheimer diseases. Less common forms of dementia include Huntington disease and frontotemporal dementia (including Pick disease), both of which occur relatively early in life (between the ages of 40 and 60). Finally, a small percentage of individuals (less than 10%) diagnosed with Alzheimer disease exhibit symptoms before age 65. This early-onset form of Alzheimer disease is distinguished from late-onset Alzheimer disease not only by the age of onset, but also by a readily identified genetic component. More specifically, dominantly inherited alleles of the *PSEN1*, *PSEN2*, and *APP* genes are now known to produce early-onset Alzheimer disease. Late-onset Alzheimer disease, on the other hand, has been linked in a less direct fashion to one variant of the *APOE* gene. The presence of a single *APOE4* allele raises the probability of developing Alzheimer disease 3 to 4 times, whereas two copies of this allele raise that probability 12 to 15 times. As such, the presence of *APOE4* can be used to assess risk, but not to definitively predict the disease.

Dyslexia and dysgraphia are learning disabilities marked by deficits in understanding or using spoken or written language. Dyscalculia is the inability to perform basic mathematical or arithmetic operations by a person who scores within the normal range of intelligence. Learning disabilities are not necessarily associated with low IQ scores, though some disabilities result in lower scores on specific portions of intelligence tests. By one estimate, 5%–15% of the general population meets the diagnostic criteria for one or more leaning disabilities. Dyslexia accounts for over half of all diagnosed cases of learning disabilities. Approximately one-third of all children diagnosed with this learning disability have at least one parent who exhibits noticeable reading problems. Four separate genes have been implicated in dyslexia: *ROBO1, EKN1, KIAA0319,* and *DCDC2.* Each of these four genes is believed to be involved in directing neural development processes, such as neuronal migration, axonal growth, and dendritic connections. The heritable nature of dyslexia has been further demonstrated by family studies showing an increased incidence in children of dyslexic parents and concordance rates of 68% for monozygotic twin pairs compared with 38% for dizygotic twins.

Review Questions and Exercises

- List ten factors that can contribute to mental retardation. Note whether each factor is associated with genetics, problems in fetal development, complications at birth, or exposure to environmental influences during early childhood.

- Compare and contrast two trisomy disorders.

- Briefly describe the phenomenon of premutation. Use features of fragile X syndrome in your description.

- Explain how microdeletion syndromes occur and list three examples of such syndromes.

- Characterize the symptoms of Rett syndrome, including the differences in effects seen in males and in females.

- At what age is Alzheimer disease considered to be of the early-onset form? Describe the differences in the genetic factors that contribute to late-onset and early-onset Alzheimer disease.

- Describe Lewy body dementia, and explain why some researchers believe that this form of dementia represents comorbid expression of two more common disorders.

- List at least three types of learning disabilities. Briefly describe the relationship between learning disabilities and measures of intelligence such as the IQ test.

FIGURE 11.2 Neuroleptic drugs reduce symptoms in many individuals diagnosed with schizophrenia. These drugs work by binding to and blocking postsynaptic dopamine receptor proteins, thereby limiting the effect of endogenously released dopamine molecules. The effectiveness of these drugs in reducing symptoms led to the dopamine hypothesis of schizophrenia, which proposes that the basis for this disorder is overactivation of the dopamine system.

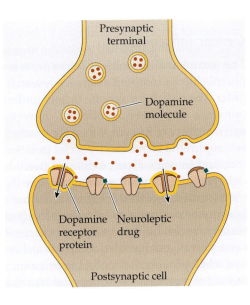

tioned in Chapter 3). However, these genes have not been at the forefront of genetics research. In explanation, it should be understood that dopamine systems are highly integrated with other neurotransmitter systems. As such, overactivation of systems producing an excititory neurotransmitter such as glutamate, for example, could in turn increase the activity of normally functioning dopamine systems, leading to behaviors indicative of dopamine overactivation (Figure 11.3).

In fact, one of the most promising candidate genes for schizophrenia, the **neuroregulin 1 (*NRG1*) gene**, located on the short arm of chromosome 8 (8p21–p22), controls production of one type of glutamate receptor protein. Specifically, Kirov and colleagues (2005) noted in a recent review of genetic contributions to schizophrenia that mutations of the *NRG1* gene have been conclusively associated with schizophrenia through a number of linkage studies. If we accept the idea that aberrations of the *NRG1* gene could contribute to abnormal behavior, it is easy to imagine a number of ways in which they could do so. For example, an allele of *NRG1* that produced hyperresponsive glutamate receptor proteins in dopamine-releasing neurons could directly promote dopamine system overactivation (Figure 11.4A). However, *NRG1* alleles that produced hyporesponsive glutamate receptors could also lead to overactivation of dopamine systems if those hyporesponsive receptors were located on cells that inhibited dopamine-releasing neurons (Figure 11.4B), or if those hyporesponsive receptors resulted in compensatory increases in glutamate release capable of stimulating other glutamate receptor subtypes on dopamine-releasing neurons (Figure 11.4C). These

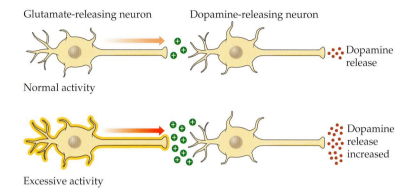

Glutamate-releasing neuron   Dopamine-releasing neuron

Dopamine
release

Normal activity

Dopamine
release
increased

Excessive activity

FIGURE 11.3 The excessive dopamine activity thought to underlie symptoms of schizophrenia could result from indirect influences of other mediating neurotransmitter systems. For example, excessive activation of an excitatory glutamate system could in turn increase output from postsynaptic dopamine-releasing cells. In this case, even though dopamine might influence the schizophrenic phenotype, the dopamine system might be unaltered by genetic mutations. Instead, genes altering the excitability of specific glutamate systems might underlie schizophrenia. Note that in this case, neuroleptic drugs would remain a viable therapeutic option for reducing symptoms caused by dopamine overactivity.

possibilities represent only three simplistic theories; given the complex nature of neural systems, other, more complex possibilities certainly exist.

The Kirov research group also concluded, based on their review of linkage studies, that the **dystrobrevin-binding protein 1 (*DTNBP1*) gene**, located on the short arm of chromosome 6 (6p22.3), is a likely candidate gene for schizophrenia. As with *NRG1*, there is no evidence to suggest that *DTNBP1* directly affects dopamine systems. Instead, *DTNBP1* is believed to play a more general role in regulating dendrite growth, density, and function. In fact, recent findings suggest that *DTNBP1* may have a more direct role in the regulation of glutamate transmission than in the regulation of dopamine transmission (Numakawa et al., 2004).

Perhaps the most intriguing finding from genetic studies of schizophrenia is that their results do not coincide directly with those of pharmacological studies. In other words, while the dopamine hypothesis suggests that schizophrenia arises directly from dysfunction of dopamine systems, that theory is not clearly supported by the genetic findings. Such conflicting results are by no means problematic, however. On the contrary, a more detailed conceptualization of the interactions between multiple neurotransmitter systems, and their combined influence on specific pathologies, may prove invaluable, not only for understanding and treating schizophrenia, but also for understanding and treating other complex polygenic disorders.

FIGURE 11.7 Autism is characterized by social and communication deficits diagnosed before the age of 3 years. Although symptoms of autism can hinder learning and education, the disorder itself does not necessarily affect cognitive abilities. Pictured here is Dr. Temple Grandin, Professor of Animal Sciences at Colorado State University. Dr. Grandin, unable to speak until after her third birthday, was diagnosed with autism. In spite of this diagnosis, Dr. Grandin earned her Ph.D. in animal sciences and is the author of over 300 scientific articles and four books, including *Animals in Translation*, which was a New York Times Best Seller.

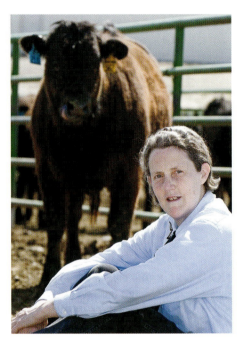

als with schizophrenia. In fact, many accounts of "childhood schizophrenia" documented through the 1800s and early 1900s appear in hindsight to have been cases of autism. The establishment of autism as a unique pathology came about in the 1940s, when at least two researchers working independently determined that this disorder was not accurately described as a variant of schizophrenia.

Even with this disorder now established as a specific psychopathology, little is known about its cause, and attempts to develop effective pharmacological treatments have proved largely unsuccessful. The severity, pervasiveness, and early onset of autism, combined with the lack of definitive findings about factors contributing to it, have made it a disorder of great interest to behavior genetics researchers.

Prevalence

In 2003, the U.S. Centers for Disease Control and Prevention published a report in which it estimated that the percentage of children with autism ranged from 0.2% to 0.6% of the general population. Perhaps more interesting was the finding that this number reflected a significant increase over previous estimates from as recently as the 1990s (Figure 11.8). One possible explanation for this in-

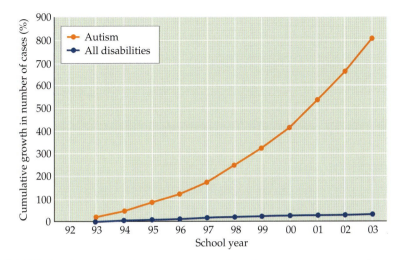

FIGURE 11.8 The number of autism diagnoses has been increasing since the 1990s. This graph, created using data from the Centers for Disease Control and Prevention and the National Center for Health Statistics illustrates the recent increase in autism diagnoses relative to those of other disabilities in individuals aged 6–22 in the United States and outlying areas (the District of Columbia, BIA schools, Puerto Rico, and the outlying areas) over the U.S. school years 1992–2003. (After Fighting Autism, www.fightingautism.org, 2008.)

crease might be that the criteria used for diagnosing autism have shifted slightly in recent years. Such a shift could move children with varying levels of mental retardation, who were formerly diagnosed with other developmental disorders, into the category of autism. For example, autism is known to occur in some individuals who suffer from Rett syndrome and fragile X syndrome, both of which are X-linked disorders (described in Chapter 10). It is possible that children with these disorders who were formerly diagnosed exclusively with mental retardation are now being diagnosed with autism. Yeargin-Allsopp and colleagues (2003) recently argued against this theory as a complete explanation. Instead, they suggested that the increasing number of children diagnosed with autism represents, at least in part, an increasing presence of this disorder. However, no researchers have yet established a possible cause for such an increase.

The prevalence of autism is greater in males than in females, with a ratio of approximately 4:1 frequently cited. It should be noted, however, that this ratio, though relatively accurate for cases of mild cognitive impairment, does not apply to cases of more severe impairment (Yeargin-Allsopp et al., 2003). More specifically, in cases of autism marked by profound cognitive impairment, the male-to-female ratio drops below 2:1. In any case, the apparent discrepancy between numbers of males and females diagnosed suggests a direct influence of genes located on sex chromosomes, an influence of reproductive hormones, or the combined influences of both of these factors. As noted above, the association of some forms of autism with X-linked cognitive disorders provides at least a partial explanation for sex differences in the numbers of individuals diagnosed.

Genetic influences

As with schizophrenia, the heterogeneous nature of autistic phenotypes has led most researchers to believe that this disorder includes several subtypes. These subtypes could result from unique combinations of several contributing genes, each responsible for mediating some distinctive underlying facet of autism. It is thus not surprising that many candidate genes, regulating a wide variety of proteins, have been implicated. A recent review of current research by Persico and Bourgeron (2006) lists 25 of the most promising candidate genes for autism and related disorders. Although these authors do provide some organization for this long list of genes, creating subcategories based on the types of proteins they influence, the complexity of the possible genetic contributions remains daunting. For example, the subcategory of candidate genes that control receptor and transporter proteins comprises seven possible genes, influencing five different neurotransmitter systems (Table 11.1).

TABLE 11.1 Candidate Genes for Autism That Encode Receptor and Transporter Proteins

GENE	LOCATION	RECEPTOR OR TRANSPORTER SYSTEM
GRIN2A	16p13	NMDA (glutamate) receptor subunit
GRIK2	6q16–q21	Kainate (glutamate) receptor subunit
SLC25A13	2q31	Aspartate–glutamate carrier
GABAR	15q12	GABA receptor subunit
SLC6A4	17p11	Serotonin transporter
OXTR	3p25–p26	Oxytocin receptor
AVPR1	12q14	Vasopressin receptor

Source: Persico and Bourgeron, 2006.

Because of the early onset of autism, researchers have been particularly interested in genes that influence proteins involved in neural development. In addition, some researchers have focused specifically on genes implicated in cognitive impairment, for two reasons. First, many syndromes known to produce mental retardation also satisfy the diagnostic criteria for **autism spectrum disorder** (**ASD**), a term used to describe syndromes containing a wide range of autistic characteristics associated with deficits in social interactions and communication skills. Second, though the hallmark symptoms of autism are difficulties with social interactions, some researchers view these deficits as impairments in the processing of information about social situations and in language interpretation. In this regard, autism could be an impairment of specific cognitive functions similar to other learning disabilities such as dyslexia.

Heritability

Interestingly, although the genetic basis for autism remains highly elusive, its inheritance patterns suggest that genetic influences on this disorder are among the strongest for any psychopathology. For example, depending on the diagnostic criteria used, concordance rates for monozygotic twin pairs range from 60% to 90%. In contrast, concordance rates for both dizygotic twins and for non-twin siblings are less than 10% (Bailey et al., 1995). The high concordance rate for monozygotic twins suggests a strong genetic contribution, whereas the very low rate for dizygotic twins indicates little or no effect of shared environment.

Folstein and Rutter (1977) were unable to find any recorded cases of a parent affected with autism who gave birth to a child with the disorder. Although such cases are likely to exist, their rarity is not surprising, considering that adults with autism seldom produce offspring. The reason for this is that, as with schizophrenia, the behaviors associated with autism tend to make these individuals less interested in engaging in sex and less desirable as sexual partners.

Eating Disorders

The two most prevalent eating disorders are anorexia nervosa and bulimia nervosa. **Anorexia nervosa**, whose name translates literally to "no appetite by nervousness," is characterized by refusal to consume enough food to maintain normal body weight, even as the affected individual experiences normal or above-normal hunger sensations. The lack of food consumption is generally motivated by an intense fear of gaining weight—a fear that is highly irrational considering the affected individual's emaciated physical state (see Figure 5.3C).

Cognitive behavior therapy may be effective in correcting some features of the disordered thought processes that underlie this irrational perception and behavior. The objective of this form of therapy is to replace maladaptive irrational thought processes with those that are more realistic and functionally adaptive. The relative success of cognitive behavior therapy suggests that a cognitive component may contribute to eating disorders in general.

Health risks associated with anorexia include renal dysfunction, cardiac arrhythmia, and **amenorrhea** (cessation of menstruation). In severe cases, death may result from kidney failure, heart failure, or malnutrition. Morbidity rates range from 10% to 20%, and only about half of all patients seeking treatment for anorexia recover completely.

Bulimia nervosa, whose name translates literally to "ox hunger by nervousness," is marked by recurrent episodes of excessive binge eating followed by inappropriate compensatory behaviors used to prevent weight gain. *DSM-IV* defines both purging and nonpurging subtypes. **Purging bulimia** is characterized by the use of vomiting, laxatives, diuretics, or enemas to reduce caloric absorption following binging periods. **Nonpurging bulimia** is characterized by excessive exercise or extended periods of fasting to lose weight gained during binging episodes. The observation that individuals diagnosed with anorexia nervosa sometimes engage in patterns of binging and purging suggests a common basis for these two disorders.

Prevalence

About 1% of individuals in the United States meet the diagnostic criteria for anorexia nervosa. The prevalence of this disorder is far greater in females, who make up about 90% of all cases. Bulimia nervosa affects slightly more individuals, with estimates of prevalence ranging from 1% to 3% in the United States. Like anorexia, bulimia affects far fewer men than women. Approximately 0.1% of males in the United States are diagnosed with bulimia.

One interesting feature of these eating disorders is that their prevalence appears to be higher in some vocational areas. High-risk occupational groups for both anorexia and bulimia are, as expected, those in which thinness is an advantageous trait. Such occupations include professional dancing, fashion modeling, and acting (Figure 11.9). Professional athletes prone to these disorders include competitive skaters, long-distance runners, and gymnasts. College sorority members are also diagnosed with eating disorders at a higher than average rate, as these social organizations frequently emphasize and reward thinness (Mehler, 1997). There has also been a suggestion that high school and college men who participate in the sport of wrestling show an elevated risk of disordered eating behaviors (Lakin et al., 1990). However, some athletes who meet the diagnos-

(A)

(B)

FIGURE 11.9 The prevalence of eating disorders is especially high in certain vocational fields. Advances in the understanding and treatment of eating disorders have prompted an increasing number of celebrities in these fields to publicly acknowledge their own experiences with these pathologies. (A) Actress Mary-Kate Olsen reported seeking treatment for anorexia nervosa. (B) Paula Abdul has revealed that she suffered from bulimia nervosa for over a decade during the early stages of her entertainment career.

tic criteria for an eating disorder (most frequently nonpurging bulimia) during their competitive season fail to meet the criteria when not actively competing in their sport (Dale and Landers, 1999).

Genetic influences

The search for candidate genes for eating disorders is in the early stages. However, findings from association and linkage studies have provided some preliminary insights that correspond nicely with earlier pharmacological data. More specifically, it has long been established that selective serotonin reuptake inhibitors are the most effective class of drugs for treating these disorders, which are, for the most part, resistant to pharmacological therapies (Vaswani and Kalra, 2004). Thus, it is not surprising that the most promising candidate genes are linked, either directly or indirectly, to serotonin (5-HT) function. In an extensive review of recent literature, Klump and Gobrogge (2005) concluded that one particular allele of the **5-HT2A gene** was most frequently linked to anorexia. The 5-HT2A gene, located on the long arm of chromosome 13 (13q14–q21), codes for the 2A subtype of the serotonin receptor protein (5-HT$_{2A}$). The same review cited evidence implicating the **5-HT2C gene** as a viable candidate gene for anorexia. This gene, located on the long arm of the X chromosome (Xq24), controls production of the 2C serotonin receptor subtype (5-HT$_{2C}$). Numerous studies assessing the contributions of other serotonin receptor subtypes have failed to show any consistent association with disordered eating behavior. These findings suggest that within the serotonin system, the influence of the 2A and 2C receptor subtypes is relatively specific. Furthermore, although anorexia is pre-

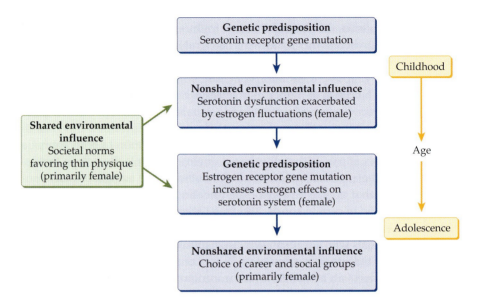

FIGURE 11.10 The theoretical model for development of eating disorders is affected by genetic, shared environmental, and nonshared environmental influences. Although genetic aberrations in serotonin receptors might affect males and females equally, these receptors might also be influenced by estrogens. Alterations in estrogen receptor genes could further increase the influence of estrogens on serotonin function. The sex-specific effects of estrogens might help to explain the preponderance of females among individuals diagnosed with these disorders. Other shared and nonshared environmental influences on the development of eating disorders also tend to be largely specific to females.

Addictive Behavior

Addiction is defined as a chronic, compulsive craving to participate in activities that evoke an emotional response. Common examples of such activities include drug use (including alcohol consumption), food consumption, gambling, and sexual behavior. As the energy expended in seeking objects of addiction increases, addictive behavior becomes incompatible with normal daily activities. Addiction has been recognized throughout history as a maladaptive psychopathology. However, because the biological basis for this behavior has remained largely unknown, addictive properties have been routinely attributed to the objects of addiction. The term "addictive drug," for example, is a common misnomer, implying that such a substance has properties that universally cause addiction. In fact, the likelihood of becoming addicted to any such drug varies widely among individuals.

Researchers generally agree that most forms of addiction represent the end result of conditioned learning. In conditioned learning an animal learns to engage in and repeat behaviors that are associated with some form of reward. In this sense, a preoccupation with pleasure-evoking behaviors (e.g., drug use, gambling) is not unusual. However, when pursuit of these behaviors continues in the face of diminished reward (e.g., drug tolerance) and/or dire consequences (e.g., failing health, financial ruin), the result of the learning process becomes counterproductive.

There is less agreement regarding the specific cognitive functions associated with normal reward seeking or the basis for the cognitive dysfunctions that drive compulsive consumption. Furthermore, consensus on what factors might predispose individuals to addiction is far from established. However, many researchers have recently begun to consider the probability of a common origin for all addictive behavioral phenotypes (see Shaffer et al., 2004). According to this theory, genetic factors combine with environmental influences to create the basis for an **addiction syndrome**. More specifically, the genetic factors produce anxiety or similar mental distress that causes the affected individual to seek relief by interacting with some element of the environment. The form that relief takes, however, depends on the individual's immediate environment. For example, an individual predisposed to addiction who lives in a region where gambling activities are highly encouraged is inclined to express a gambling addiction phenotype. On the other hand, a person with similar predisposing factors who lives in a region where drug use, but not gambling, is encouraged will not be inclined to become addicted to gambling; instead, a drug addiction phenotype is likely to be expressed. If one accepts the theory that a single cause (or set of causes) underlies all addictive behavior, then it is reasonable to consider common biological underpinnings, such as genetic influences.

Prevalence

Estimates of the numbers of individuals who are affected by addictions are highly speculative. To begin with, the nature of some addictions means that they are less frequently reported than others. For example, food and sex addictions are generally far less disruptive to normal daily activities than are addictions to alcohol and other psychoactive drugs. In addition, though a gambling addiction may be highly maladaptive in a person who earns only enough money to subsist, an identical or even more severe form of this pathology may be less problematic for a wealthy individual (**Figure 11.11**). In fact, wealth can reduce some of the problematic features of all types of addiction, including addiction to psychoactive drugs. The severe health and (in some cases) legal consequences associated with alcohol and other drug addictions, however, are largely universal across socioeconomic classes.

FIGURE 11.11 The extent to which addiction disrupts an individual's normal daily activities depends of the form of the addiction and on the individual's circumstances. In his autobiography, golfer John Daly acknowledged that a 12-year gambling addiction cost him millions of dollars. His lucrative income as a world-class professional golfer allowed him to maintain a level of addictive behavior that would have been extremely disabling to most individuals.

In spite of these limitations, some numbers may be gleaned from the literature to provide rough estimates of the prevalence of addictions. Nicotine is by far the greatest object of addiction in the United States, where about 15% of the general population reports compulsive use of this drug. About 4% of U.S. residents meet the *DSM-IV* criteria for alcohol dependence (Table 11.2), which clearly reflect addictive behavior. Estimates of the number of U.S. residents addicted to illegal psychoactive drugs are wide ranging, but it is probably somewhat smaller than the number addicted to alcohol. Addictions to other psychoactive substances may be possible as well (Box 11.1). Estimates of the

---

TABLE 11.2 *DSM-IV* Criteria for Alcohol Dependence

1. Tolerance (e.g., needing more alcohol to become intoxicated)
2. Withdrawal symptoms or use of alcohol to relieve or avoid those symptoms
3. Alcohol use for longer periods than intended
4. Desire and/or unsuccessful efforts to cut down or control alcohol use
5. Considerable time spent obtaining or using alcohol, or recovering from its effects
6. Important social, work, or recreational activities given up because of use
7. Continued use of alcohol despite knowledge of problems caused by or aggravated by use

Source: Adapted from *DSM-IV*, 1994.
Note: A person is defined as being dependent on alcohol if he or she reports three or more of these symptoms in the past year.

percentage of U.S. residents addicted to gambling also vary widely, but most estimates are in the range of 1% to 3%. Addictions to food and sexual activity are assumed to be widely underreported, making estimates difficult. It is, however, reasonable to believe that the prevalences of these addictive behaviors are within or below the range for gambling addiction. Finally, it should be kept in mind that comorbid addictions are frequently reported, with the combination of nicotine and alcohol addiction being most common.

BOX 11.1　Is the Java Jive a Psychopathology?

I love coffee, I love tea, I love the java jive and it loves me.

From "Java Jive," words and music by Milton Drake and Ben Oakland, 1940

Why is there no formal recognition of caffeine addiction? Nicotine and alcohol are both legal to purchase in most countries around the world. Both have psychoactive qualities, and both have the potential for development of dependence and addiction among users. Interestingly, these same qualities are characteristic of caffeine, yet it is often not classified as an addictive drug. In fact, *DSM-IV* recognizes both alcohol and nicotine dependence, but not caffeine dependence. Consider for a moment that meeting the following four criteria qualifies nicotine and alcohol users for a diagnosis of drug dependence:

- Continued use despite knowledge of a persistent or recurrent physical or psychological problem that is likely to have been caused or exacerbated by the substance

- Persistent desire or unsuccessful efforts to cut down or control substance use

- Characteristic withdrawal syndrome or use of the substance to relieve or avoid withdrawal symptoms

- Tolerance, as defined by a need for markedly increased amounts of the substance to achieve the desired effect, or a markedly diminished effect with continued use of the same amount of the substance

A large number of individuals who regularly consume caffeine meet these four criteria, so why no addiction diagnosis?

In the case of caffeine, it appears that an addiction exists, but may simply not yet be officially recognized. Although *DSM-IV* certainly does address the problematic features of caffeine consumption (caffeine-induced anxiety disorder, caffeine-induced sleep disorder, caffeine intoxication, caffeine withdrawal), the failure to include its chronic use as a drug dependence disorder seems to be based on a lack of supporting research rather than a disbelief that addiction to this drug is possible. Evidence for that possibility is growing, however, and there is speculation that caffeine dependence may appear in a future edition of *DSM*.

Consider for a moment the effect of recognizing caffeine addiction on the public perception of the prevalence of addictive behavior. It is estimated that about 80% of adults and adolescents in North America consume caffeine on a regular basis. There are no age restrictions on the consumption of this drug. In addition to commonly recognized sources, such as coffee and tea, caffeine is found in foods (including candy), soft drinks, energy drinks, and in pure form in tablets. Could it be that over half of the adults in North America express an addictive phenotype to some degree? It is certainly a question worth mulling over while drinking your morning coffee.

Genetic influences

The majority of genetic research into addictive behaviors has focused on genes that regulate dopamine systems. Neuropharmacological research indicates that all drugs with addictive potential enhance dopamine transmission, either directly or indirectly. Furthermore, there is a notable correlation between the level of dopamine increase produced by a particular drug and its potential as an addictive substance. Other objects of addiction similarly increase dopamine output. Food consumption, sexual behavior, gambling, and even running and other forms of aerobic exercise (which have limited addictive potential) all increase dopamine concentrations in the mesolimbic system. These findings have led to the theory that the mesolimbic dopamine system constitutes an **endogenous reward system**. Theoretically, the evolution of such a reward system would be advantageous in the sense that behaviors essential to survival and procreation—particularly eating and sex—would be reinforced. However, a lack of specificity in the reward system that allowed it to respond to a range of other stimuli would leave open the possibility of the individual's developing an affinity for many behaviors capable of activating this system. In that case, compulsive engagement in maladaptive behaviors that stimulate the reward system would accurately describe addictive behavior.

With the concept of a dopamine-mediated reward system in mind, researchers have identified several candidate genes for addiction, all of which either directly or indirectly affect dopamine activity. The **DAT1 gene**, located on the short arm of chromosome 5 (5p15.3), regulates neuronal dopamine transporter proteins that are responsible for dopamine reuptake. Recall from Chapter 3 that proteins embedded in the presynaptic membrane selectively remove neurotransmitter molecules from the synapse, reducing neurochemical communication while recycling these chemicals back into the presynaptic axon terminal for future use (Figure 11.12). Dopamine transporter proteins are directly affected by several drugs that have a high degree of addictive potential, the foremost being cocaine and amphetamines. Further evidence implicating dopamine transporter proteins in the development of addictive behaviors is the finding of specific *DAT1* alleles among some individuals who exhibit a high level of vulnerability to drug use and addiction (see Hurd, 2006 for review).

Addictive behavior is frequently comorbid with antisocial personality traits, a topic that will be discussed further in Chapter 12. One possible explanation for this comorbidity is a common *DAT* gene mutation underlying both antisocial and addictive behavior. Gerra and colleagues (2005) have suggested that one particular *DAT* allele appears to contribute to antisocial behavior, and that expression of this phenotype promotes addictive behaviors. Their theory is based on their finding of this allele in antisocial drug addicts, but not in addicts who were not diagnosed with antisocial personality disorder.

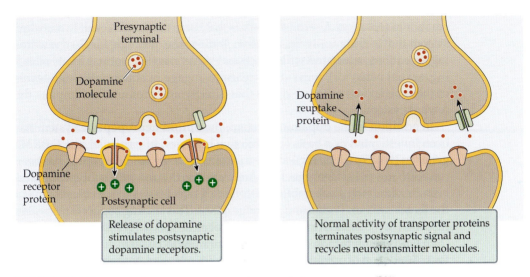

Release of dopamine stimulates postsynaptic dopamine receptors.

Normal activity of transporter proteins terminates postsynaptic signal and recycles neurotransmitter molecules.

FIGURE 11.12 The *DAT1* gene controls production of dopamine transporter proteins that are responsible for dopamine reuptake. These proteins, which are embedded in the presynaptic axon terminal membrane, have the primary function of returning dopamine molecules to the terminal after they have been released into the synapse. This function is important for deactivating dopamine-mediated signals and for recycling dopamine molecules, making them available for rapid rerelease with subsequent neuronal activation. *DAT1* gene variations have been identified in some individuals who are particularly vulnerable to drug use and addiction.

If alterations of dopamine transporter protein function could predispose an individual to addictive behaviors, then it is reasonable to assume that similar dopamine system dysfunctions caused by alterations of other proteins could have similar behavioral effects. The **DRD2 gene**, located on the long arm of chromosome 11 (11q22–q23), regulates production of the D_2 dopamine receptor subtype. One mutation of this gene has also been linked to addictive behaviors (Noble, 2000). The shared phenotype of these two gene mutations (of *DAT1* and *DRD2*) suggests that they may have a common effect on dopamine transmission, most likely in the mesolimbic system.

It is currently unclear how alterations of dopamine transporter proteins or D_2 dopamine receptor proteins could increase the likelihood of addiction. One theory is that abnormal structures or quantities of these proteins create a reward system that is hyperresponsive to stimulation by the object of addiction (i.e., drugs). In this case, the reward sensation that is cognitively associated with the stimulus becomes highly reinforcing. Such an association could lead to an obsession with obtaining the stimulus that produces the exaggerated

Although heritability studies indicate a genetic influence on addictive behaviors, environmental influences appear to have a strong effect as well. In particular, shared environmental factors contribute to substance abuse patterns, especially when experienced during adolescence. Han and colleagues (1999) note that these shared influences could include neighborhood atmosphere, parental attitude toward substance use, school environment, and shared peer groups. It is noteworthy that these influences are reduced or eliminated before adulthood. Although exposure to these shared environmental factors may be restricted to a limited period during adolescence, it has been suggested that initiation of abuse patterns stimulated by these factors often precedes a more enduring addictive behavioral phenotype seen in adulthood.

Finally, it is worth mentioning that a high percentage of individuals diagnosed with other psychopathologies, including schizophrenia and mood, anxiety, and personality disorders, also exhibit addictive behaviors. The finding of such a high degree of comorbidity with other seemingly unrelated disorders underscores the importance of advancing research in addictive behaviors. It would appear that a greater understanding of genetic influences on dopamine systems in particular may be useful in elucidating the organic basis for numerous disorders.

Chapter Summary

Schizophrenia affects about 1% of the adult general population and ranks among the top ten causes of reported disabilities in industrialized countries. Only about 1 in 4 individuals diagnosed with schizophrenia are capable of living independently. Although schizophrenia affects approximately equal numbers of men and women, the age of onset is generally earlier for men. Some researchers believe that, rather than a single disorder, schizophrenia may represent a group of syndromes with a common feature of severe psychosis marked by disorganized thought processes and a loss of touch with reality.

Neuroleptics are drugs that decrease dopamine transmission and reduce some symptoms of schizophrenia. The therapeutic effectiveness of these drugs suggests that dopamine overactivation, particularly in the mesolimbic dopamine system, may cause some symptoms of schizophrenia. However, the search for candidate genes that regulate dopamine transmission has not been highly productive. Genes more reliably linked to schizophrenia include the *NRG2* gene, which controls production of glutamate receptors, and the *DTNBP2* gene, which has a general role in regulating the development and function of dendrites.

The heterogeneous nature of schizophrenic symptoms suggests the involvement of numerous genes. Both pedigree studies and twin studies support the theory of a genetic contribution. A concordance of 48% for monozygotic twin pairs compared with 17% for dizygotic twin pairs shows a genetic influence on this pathology. Nonshared environmental factors are also believed to contribute to schizophrenia, with possible factors including complications at birth, prenatal drug exposure, and exposure to influenza viruses during fetal development.

Although autism is widely recognized as a specific psychopathology, little is known about its cause, and pharmacological treatments are largely ineffective. It was recently estimated that 0.2%–0.6% of individuals in the general population are affected by autism. Interestingly, this number reflects a significant increase when compared with estimates from a decade earlier. There has been some speculation that a shift in diagnostic criteria underlies this increase, but at least one group of researchers has argued against this theory and suggested that some undiscovered cause has influenced the recent increase in autism diagnoses.

In general, autism is four times more prevalent in males than in females. However, that ratio is reduced to 2:1 in cases of autism characterized by profound cognitive impairment. This disproportionate sex ratio suggests that the genes involved may be located on sex chromosomes or may influence reproductive hormones. One recent review identified 25 possible candidate genes for autism, controlling a wide range of proteins associated with several different neurotransmitter systems. Genes that regulate neural development and cognitive functions have generated the most interest. Although a specific genetic basis for autism has not been identified, its patterns of inheritance point to a very strong genetic influence.

Anorexia nervosa is a refusal to consume enough food to maintain normal body weight, generally motivated by an intense, irrational fear of gaining weight. Bulimia nervosa is characterized by recurrent episodes of excessive binge eating followed by purging, excessive exercise, or extended periods of fasting. Individuals diagnosed with anorexia nervosa sometimes engage in patterns of binging and purging, suggesting a common basis for these two disorders. About 1% of individuals in the United States meet the diagnostic criteria for anorexia, with females accounting for approximately 90% of all cases. Bulimia affects 1%–3% of the U.S. population, with females again accounting for more than 90% of all cases. Rates of anorexia and bulimia are elevated in some occupational and social groups that encourage a thin physique.

Research has linked the *5-HT2A* and 5-*HT2C* serotonin receptor genes, as well as the estrogen receptor beta (*ER*) gene, to anorexia. Concordance rates greater than 50% in monozygotic twin pairs and less than 10% in dizygotic twins suggest a strong genetic influence on anorexia. For bulimia, the concordance rate for monozygotic twins is about 23%, whereas that for dizygotic twins is just below 9%. Rates of all types of eating disorders are significantly higher than average among first-degree relatives of individuals with both anorexia and bulimia. Anorexia and bulimia also appear to be influenced by some shared environmental factors, most notably the influence of living in cultures where a thin physique is considered sexually appealing.

Addiction can be defined as a chronic, compulsive craving to participate in activities that evoke an emotional response. About 15% of the population reports compulsive use of nicotine. About 4% meet the *DSM-IV* criteria for alcohol dependence, and estimates of gambling addiction range from 1% to 3%. Most genetic research on addictive behaviors has focused on genes that regulate dopamine systems. Candidate genes include the *DAT1* gene, which controls transporter proteins responsible for dopamine reuptake, and the *DRD2* gene, which regulates production of the D_2 dopamine receptor subtype. Addictive behavior is frequently comorbid with antisocial personality disorder, and evidence suggests that one *DAT* allele may contribute to the expression of both traits in some individuals.

Addiction could result from gene abnormalities that create a hyperresponsive mesolimbic dopamine reward system. In this case, behaviors that stimulated an overly responsive reward system (including psychoactive drug use, gambling, or sexual activity) could, in turn, be highly reinforced, leading to an obsession with obtaining the stimulus associated with dopamine activation. A second possibility is that gene aberrations could result in a chronically underactive reward system. According to this theory, a lack of rewarding stimulation associated with normal levels of functional behavior could cause individuals to engage in abnormal (excessive) behaviors to activate the reward system.

Heredity studies have revealed correlations of .40 for excessive alcohol use and .38 for frequent recreational drug use between monozygotic twins, compared with rates of .29 for alcohol and .22 for drug use in dizygotic twins. First-degree relatives of alcohol abusers were reportedly twice as likely to be alcohol abusers, and first-degree relatives of drug abusers were as much as 8 times more likely to abuse drugs, than were unrelated individuals. Together, these findings strongly suggest a genetic contribution to addictive traits. However, environmental influences, particularly shared environmental factors, also have an influence on addictive behavior. Factors that may be influential include neighborhood atmosphere, parental attitude toward substance use, school environment, and shared peer groups. All of these environmental factors appear to be most influential when experienced during childhood or adolescence.

Review Questions and Exercises

- Explain how recent genetic research has been integrated into the dopamine hypothesis of schizophrenia.

- List at least three environmental factors that have been implicated as possibly contributing to the development of schizophrenia.

- Give at least two examples of how alterations in glutamate system function could result in excess dopamine activity, such as that thought to occur in schizophrenia

- Compare and contrast autism with schizophrenia and with mental retardation.

- Describe at least two possible factors that could contribute to differences in prevalence rates of autism between males and females.

- Briefly summarize possible genetic and environmental contributions to anorexia.

- Explain why genetics research into addictive behavior has focused primarily on genes that regulate dopamine system activity.

- Briefly describe research findings that suggest a link between addictive behavior and antisocial personality disorder.

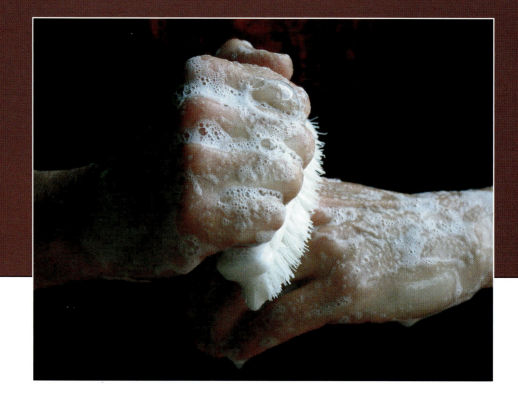

12

Disorders of Mood, Anxiety, and Personality

Chapter 9 discussed genetic influences on the normal range of intelligence and behavior. Chapters 10 and 11 followed with discussions of genetic influences on primary disorders of cognition and of other pathologies that may have a basis in cognitive dysfunction. This chapter will discuss abnormal personality traits and closely associated mood and anxiety disorders. In some ways, the methods used for evaluating mood, anxiety, and personality parallel those used for evaluating intelligence. Specifically, quantifiable tests used to evaluate these traits generate a relatively normal distribution of scores for the general population. Scores that fall above or below the normal range typically represent maladaptive mood, anxiety, or personality characteristics. It appears, however, that more individuals fall outside the normal range than would be predicted by the normal distribution. Another difference between studies of intelligence and studies of mood, anxiety, and personality is the number of distinct trait subcategories. Although intelligence may encompass numerous cognitive abilities, it is nonetheless considered a fairly homogeneous trait. Mood, anxiety, and personality, on the other hand, can each be subdivided into distinct phenotypes. It is no surprise that such subcategories imply a greater range of possible genetic factors acting on these traits. In spite of these complexities, our understanding of genetic contributions to and hereditary patterns of several major mood, anxiety, and personality disorders has increased considerably in recent years.

Before we begin this discussion of mood, anxiety, and personality disorders, it should be recognized that in many cases, these disorders are comorbid. Furthermore, distinct disorders in these categories frequently share similar symptoms. For example, there is pronounced overlap in the diagnostic criteria for generalized anxiety disorder and major depression. In fact, some researchers believe that these two disorders may be different manifestations of a common physiological dysfunction. This theory is supported, at least in part, by the observation that in some cases, successful treatment for generalized anxiety disorder and major depression can be achieved with the same drug(s). Such diagnostic overlap has the potential to add clarity to a larger picture of shared genetic and neurological contributions to mood, anxiety, and personality. In particular, if a common physiological basis for distinct disorders is found, that finding could suggest a common genetic cause. If a common genetic cause is established, researchers could then begin to assess the effects of environmental factors, or interactive effects of other genes, that may ultimately be responsible for creating the phenotypic variations associated with the pathology.

Mood Disorders

Mood disorders are characterized by a prominent and persistent mood disturbance that disrupts normal daily activities. Both **major depression** and **dysthymia** (chronic low-level depression) are marked by prolonged depressed emotional states. Such emotional states are usually characterized by general apathy, feelings of worthlessness or despair, periods of insomnia, loss of appetite, and lack of sexual drive.

Bipolar disorder, formerly referred to as **manic depression**, causes mood swings that cycle from bouts of depression to periods of mania. During these cycles, symptoms of major depression may persist for several weeks or months, followed by generally shorter periods of mania. Symptoms of the manic phase include increased physical and mental activity, accelerated thought patterns, rapid speech, and impulsiveness. Manic symptoms become disruptive when they lead to difficulty in maintaining concentration, irritability, and reckless or aggressive behavior. Although some researchers contend that **unipolar mania** is a distinct disorder, mania in the absence of intermittent depressive episodes is rare and is not recognized as a mood disorder by *DSM-IV*.

Prevalence

A recent survey found that 11.5% of U.S. residents between the ages of 17 and 39 met the diagnostic criteria for at least one mood disorder (Jonas et al., 2003). Major depression was reported by the greatest percentage of individuals (8.6%),

(A)

(B)

FIGURE 12.1 Mood disorders may affect individuals from all walks of life. (A) A number of highly successful actors and (interestingly) comedians have experienced clinical depression. Among the most prominent actor/comedians to report suffering from depression is two-time Golden Globe winner Jim Carrey. (B) Although lethargy and apathy are primary characteristic of depression, some outstanding athletes have overcome such symptoms to excel in their sports. Monica Seles, for example, rebounded from a 2 year bout of depression to win the Canadian Open in 1995 followed by the Australian Open in 1996.

followed by dysthymia (6.2%) and bipolar disorder (1.6%) (Figure 12.1). The discrepancy between the percentage reporting at least one mood disorder and the sum of the percentages for the individual mood disorders reflects the substantial number of respondents who met the diagnostic criteria for more than one disorder (Figure 12.2).

The prevalence of these disorders is probably not particularly surprising. What may be more interesting are the correlational relationships that emerged from the survey. For example, Jonas and colleagues (2003) found more frequent reports of mood disorders in individuals with less education, lower income, and poorer health. Specific factors associated with an increased likelihood of reporting a mood disorder were separation or divorce, cigarette smoking, asthma, and hypertension. These data raise two questions of interest to behavior genetics researchers: First, does a causal relationship exist between such factors and mood disorders? Second, if a causal relationship does exist, which factor is the cause and which is the effect? If, for example, smoking and hypertension increase the likelihood of mood disorders, then knowledge of how genes influence nicotine use and blood pressure could be useful in the study of mood disorders. If, on the other hand, mood disorders increase the likelihood of nicotine

FIGURE 12.2 Approximately 11.5% of U.S. residents between the ages of 17 and 39 meet the diagnostic criteria for at least one mood disorder. Major depression is the most prevalent of these disorders, affecting 8.6%, followed by dysthymia (6.2%) and bipolar disorder (1.6%). Some individuals meet the criteria for more than one mood disorder. (After Jonas et al., 2003.)

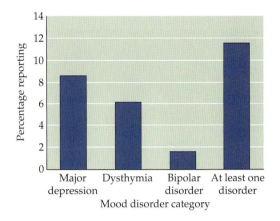

use or hypertension, then the discovery of a genetic basis for mood disorders could be useful for better understanding the basis for drug addiction and chronic high blood pressure. Finally, it is possible that these correlations reflect genetic influences that are common to both mood disorders and other dysfunctional or maladaptive traits. In this case, any genetic discoveries would be useful for all related research.

Another interesting statistic regarding mood disorders is that approximately 2 out of every 3 cases are reported by women. Although the reason for this sex-based discrepancy is unclear, there are several possible explanations. One possibility is that specific sex hormones may have differential actions on neurological functions associated with mood. A second, related possibility is that sex differences in the function of the **hypothalamic–pituitary–adrenal (HPA) axis** underlie a variety of sex differences in behavioral expression, including mood. The HPA axis influences concentrations of numerous hormones and is highly responsive to external stimuli (**Figure 12.3**). In particular, chronically elevated concentrations of the adrenal stress hormone **cortisol** are widely regarded as a precipitating factor in several mood and anxiety disorders (for a recent review, see Brown et al., 2004). A third possible reason for sex differences in reports of mood disorders may have a psychosocial, rather than a physiological, basis. Some researchers and clinicians believe that males in most cultures are less inclined to report symptoms of mood disorders than are females (see Amenson and Lewinsohn, 1981). If that is the case, the obvious result would be a high female-to-male ratio for reported mood disorders that is not influenced by biological factors. The other equally important effect of a failure by males to report symptoms would be a notable underestimation of the already high numbers of affected individuals.

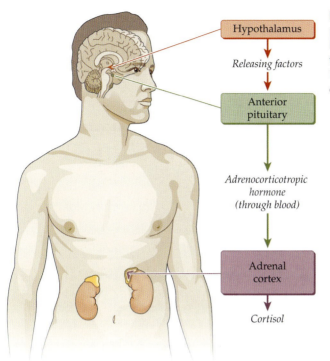

FIGURE 12.3 The hypothalamic–pituitary–adrenal (HPA) axis, a highly integrated neuroendocrine system that regulates responses to stress, has important influences on mood and anxiety.

Genetic influences

The finding that mood disorders affect more than 1 in 10 individuals in the United States underscores the urgency of discovering their causes so that effective treatments can be developed. In addition, based on the correlational data cited above, one could hope that improved treatment of these disorders would have concomitant beneficial effects on physical health, education, income potential, and other related factors. It is thus not surprising that a great deal of research effort has been directed at locating candidate genes associated with these mood disorders.

Prior to investigations focused on the genetic basis of major depression and dysthymia, researchers were aware of the fact that drugs designed to enhance serotonin neurotransmission were particularly effective in treating symptoms of these disorders. This general understanding of the neurochemical underpinnings of depression prompted behavior genetics researchers to examine the influence of genes known to control serotonin transmission. Of particular interest is the **5-HT transporter (5-HTT) gene** located on the long arm of chro-

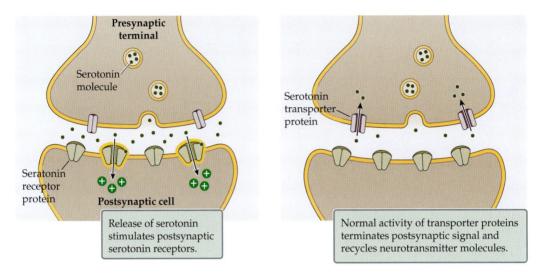

FIGURE 12.4 The *5-HTT* gene controls production of serotonin transporter proteins. These proteins are embedded in the presynaptic axon terminal membrane and have the primary function of returning serotonin molecules to the terminal after they have been released into the synapse. This function is important for deactivating serotonin-mediated signals and for recycling serotonin molecules, making them available for rapid rerelease with subsequent neuronal activation. Dysfunction of this reuptake process has been implicated as a possible cause of mood disorders.

mosome 17 (17q11.2). The *5-HTT* gene is known to regulate the structure and function of serotonin transporter proteins. Just as dopamine transporter proteins (discussed in Chapter 11) regulate synaptic reuptake of dopamine, serotonin transporter proteins are responsible for the reuptake of serotonin released into the synapse (Figure 12.4). Selective serotonin reuptake inhibitors (SSRIs) are a particularly effective class of antidepressant drugs. As described in Chapter 3, they block the activity of the transporter proteins, leading to a normalization of mood in many clinically depressed individuals. Caspi and colleagues (2003) found that the presence of one polymorphic region linked to the *5-HTT* gene (*5-HTTLPR*) was directly associated with depressive symptoms, diagnosed depression, and suicide. Moreover, these researchers found that individuals with two copies of this particular allele were more inclined to exhibit these traits than were individuals with a single copy. Finally, this allele was associated with depression only in individuals who reported stressful life events that, by themselves, did not typically evoke depression. This last point exemplifies the complex interactions between genes and environment that result in phenotypic expression, particularly for quantitative traits.

TABLE 12.1 Modeling of Symptoms of Major Depression in Mice

SYMPTOM IN HUMANS	SYMPTOM AS MODELED IN MICE
Diminished interest or pleasure in everyday activities (anhedonia)	Reduced self-stimulation of brain reward centers; reduced responding for positive reward (for example, sucrose); increased incidence of social withdrawal
Large change in appetite or body weight	Abnormal reduction in body weight after exposure to chronic stress
Insomnia or excessive sleeping	Abnormal sleep architecture (measured using electroencephalography)
Psychomotor agitation or slowness of movement	Difficulty in handling; alterations in various measures of locomotor activity and motor function
Fatigue or loss of energy	Reduced activity in home cage; reductions in treadmill or running wheel activity and nest building; reductions in waking electroencephalogram activity
Indecisiveness or diminished ability to think or concentrate	Deficits in working and spatial memory and impaired sustained attention
Difficulty performing even minor tasks, leading to poor personal hygiene	Poor coat condition during chronic mild stress
Recurrent thoughts of death or suicide	Cannot be modeled
Feelings of worthlessness or excessive or inappropriate guilt	Cannot be modeled

Source: Cryan and Holmes, 2005.

Researchers have also explored the possibility that genes affecting other features of serotonin transmission could contribute to major depression and dysthymia. Svenningsson et al. (2006), for example, recently found that the **calpactin I light chain protein**, also known as the **p11 protein**, was underexpressed in postmortem brain tissue from individuals diagnosed with depression. The p11 protein is believed to have a primary function of localizing other proteins on the neuronal cell surface, including serotonin receptor proteins. In mice in which a gene for p11 protein production has been knocked out, a depression-like phenotype emerges (Svenningsson et al., 2006) (Table 12.1). In humans, p11 protein production is controlled by the *CLP11* gene, located on the long arm of chromosome 1 (1q21) (Harder et al., 1992). Taken together, these findings suggest that underexpression of the p11 protein, possibly caused by a mutation of the *CLP11* gene, may contribute to serotonin receptor dysfunction and subsequent symptoms of depression.

(A)

(B)

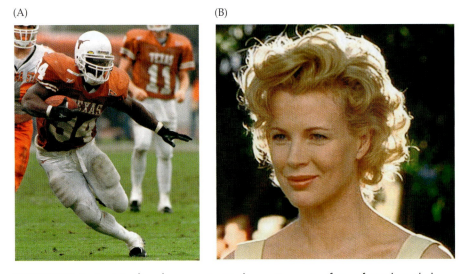

FIGURE 12.6 Anxiety disorders are among the most common form of psychopathology, affecting nearly 1 in 6 U.S. residents. Although such disorders may disrupt normal daily activities, they do not directly affect intelligence, creativity, or athletic ability. (A) Ricky Williams won the most coveted college football award, the Heisman Trophy, in 1998, at a time when he also reportedly suffered from severe social phobia. (B) Kim Basinger, who has reported suffering from both panic disorder and agoraphobia, achieved the highest honor in motion picture acting when she received an Academy Award for Best Supporting Actress.

Genetic influences

Little is currently known about which genes may directly contribute to anxiety disorders. Most antianxiety drugs, also called anxiolytics, act as agonists at GABA receptor sites, producing a general slowing of neuronal transmission, as described in Chapter 3. Although these drugs effectively treat many forms of anxiety, no consistent findings of problems with genetic regulation of GABA neurotransmission systems have been reported.

Oxidative stress results when normal cellular metabolic processes are disrupted, leading to the formation of free radicals that can ultimately damage cells. Oxidative stress has been implicated as an underlying factor in many neurological and psychiatric diseases, including anxiety disorders (Kuloglu, 2002). With this in mind, one group of researchers recently assessed differences in the genomes of six inbred strains of mice that differed greatly in behavioral expressions of anxiety (Hovatta et al., 2005). This cross-strain analysis produced 17 candidate genes for anxiety. Hovatta and colleagues found that the two most promising candidate genes in their assessment regulated enzymes within the same metabolic pathway. The *Glo1* **gene** regulates production of the enzyme

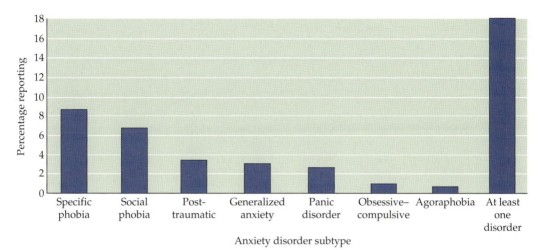

FIGURE 12.7 Percentages of U.S. residents who meet the diagnostic criteria for anxiety disorders. Approximately 18% of U.S. residents over the age of 18 meet the diagnostic criteria for at least one anxiety disorder. As with mood disorders, comorbidity is common: individuals reporting one anxiety disorder subtype frequently meet the diagnostic criteria for a second subtype as well.

glyoxalase 1, whereas the *Gsr* **gene** regulates production of the enzyme **glutathione reductase 1**. One theory based on these findings is that oxidative stress resulting from a metabolic imbalance, perhaps in this particular pathway, could contribute to anxiety disorders. Hovatta and colleagues also found that transgenic mouse models designed to overexpress glyoxalase and glutathione reductase 1 showed significant increases in behavioral expressions of anxiety. The next step in delineating these intriguing findings will be to identify genes in humans that are homologous to those influencing anxiety in mice. Once they are located, researchers can then determine whether particular alleles of these genes exist in individuals suffering from specific anxiety disorders.

Although the search for human alleles that may produce anxiety is just beginning, at least one candidate gene has generated some interest. Zuchner and colleagues (2006) recently found that two mutations of the **SLITRK1 gene** were linked to **trichotillomania**, a form of obsessive hair pulling. *SLITRK1* is located on the long arm of chromosome 13 (13q31) and controls production of one type of **Slit protein**. This family of proteins has a role in guiding neuronal migration, promoting neuronal outgrowth, and regulating the formation of connections among neurons. Mutations of the *SLITRK1* gene had been identified previously as a possible factor contributing to the development of **Tourette syndrome** (Abelson et al., 2005), which is characterized by uncontrollable tics and vocalizations. Tourette syndrome is frequently comorbid with obsessive–

compulsive disorder, leading some researchers to believe that a common organic basis may underlie both disorders.

At least one other gene has been linked to obsessive–compulsive disorder. The *hSERT* **gene**, located on the long arm of chromosome 17 (17q11.1–q12), controls production of one type of serotonin transporter protein. Ozaki and colleagues (2003) recently found that mutations of this gene were linked to obsessive–compulsive symptoms. Moreover, these researchers found that in some cases in which multiple mutations occurred in the chromosome region containing this gene, more severe symptoms of obsessive–compulsive disorder were typically reported. Taken together, these findings strongly implicate the *hSERT* gene as a contributing factor to this anxiety disorder subtype.

The finding that a serotonin transporter gene may contribute to an anxiety disorder is not surprising. Recall from the beginning of this chapter the statement that mood and anxiety disorders share many common symptoms. In fact, some researchers believe that depression and general anxiety disorder may represent different phenotypes generated by common alleles. The differences in phenotypic expression, in these cases, may simply result from the presence or absence of certain environmental factors or additional contributing genes. Table 12.3 lists and summarizes information for some of the genes currently implicated in the development of anxiety disorders.

TABLE 12.3 Some Genes Currently Implicated in Anxiety Disorders

GENE (LOCATION)	ASSOCIATED PROTEIN	DISORDER	NOTES
Glo1 (6p11)	Glyoxalase 1	Generalized anxiety?	Researched primarily in rodent models of anxiety
Gsr (8p12–p21)	Glutathione reductase 1	Generalized anxiety?	Researched primarily in rodent models of anxiety
SLITKR1 (13q31)	Slit protein	Trichotillomania; possibly obsessive–compulsive disorder	Mutations of this gene are seen in some cases of Tourette syndrome
hSERT (17q11.1–q12)	Serotonin transport protein	Obsessive–compulsive disorder	Multiple mutations of this region are correlated with more severe symptoms

Heritability

As with mood disorders, heritability studies of anxiety disorders suggest a genetic contribution to all subtypes. In their recent review of heritability studies, Merikangas and Low (2005) noted that offspring of parents with anxiety disorders have an increased risk of developing such disorders themselves. However, the anxiety disorder subtype expressed by the offspring frequently differs from that expressed by the parent(s). This finding contrasts with data from heritability studies of mood disorders, in which the parental phenotype is often predictive of the child's disorder. Merikangas and Low also found that panic disorder showed the highest degree of heritability among the subtypes studied. This finding coincides nicely with the earlier findings of Perna and colleagues (1997), who reported a panic disorder concordance rate of 73% for monozygotic twin pairs and a lack of any significant concordance rate for dizygotic twins.

Although few candidate genes have yet been identified that might be analyzed in pedigree studies, our increasing understanding of the human genome is likely to allow human linkage studies designed for this purpose in the near future. It is conceivable that data will soon be available to establish not only contributing genes, but also the typical inheritance patterns of those genes and their relative contributions to a wide array of anxiety disorder phenotypes.

Like other polygenic traits, anxiety disorders are thought to be highly susceptible to the influence of environmental factors. As with mood disorders, stress is considered a primary nonshared environmental factor contributing to the development of the pathology. In the case of anxiety disorders, stress may be an even more intuitive environmental trigger when we consider that the physiological arousal associated with any given stressor is similar to the arousal reported in these disorders. Perhaps the most obvious example of environmental stress contributing to an anxiety disorder is seen in post-traumatic stress disorder. In this case, an individual's experience of a particularly stressful life event causes subsequent episodes of anxiety directly linked to recall of that specific arousing and unpleasant experience.

Another less direct example of the influence of stress is seen in the positive correlation between age and reports of anxiety disorders. More specifically, reports of panic attacks and generalized anxiety disorder are particularly common in the elderly. This correlation is believed to be caused in large part by increasing stress associated with real or perceived degeneration of health. Such chronic physical stressors are obvious precursors to development of anxiety disorders in those who are genetically predisposed.

Some researchers note that shared environmental factors—most notably the effects of parenting style on children—could influence later development of anx-

---

TABLE 12.4 *DSM-IV* Criteria for Antisocial Personality Disorder

---

1. Failure to conform to social norms with respect to lawful behaviors, as indicated by repeatedly performing acts that are grounds for arrest
2. Deceitfulness, as indicated by repeated lying, use of aliases, or conning others for personal profit or pleasure
3. Impulsivity or failure to plan ahead
4. Irritability and aggressiveness, as indicated by repeated physical fights or assaults
5. Reckless disregard for safety of self or others
6. Consistent irresponsibility, as indicated by repeated failure to sustain consistent work behavior or honor financial obligations
7. Lack of remorse, as indicated by being indifferent to or rationalizing having hurt, mistreated, or stolen from another

---

Source: Adapted from *DSM-IV*, 1994.

ity disorders, antisocial personality disorder has unique characteristics that suggest it is not a mild form of a more severe psychopathology, but rather a unique phenotype (Figure 12.9). Second, estimates of the prevalence of antisocial personality disorder among prison inmates range from 60% to 90%, implying a strong correlation with criminal behavior (Box 12.1). Third, statistics show that this disorder is diagnosed three times more often in men than in women, suggesting an underlying genetic factor that is sex-linked or responsive to sex-specific hormones.

FIGURE 12.9 Christian Longo, convicted of killing his wife Mary Jane and their three young children in 2001, was later described as a prototypical sociopath who was able to present himself to others as an average, or even a model, citizen. Dr. Martha Stout gives an excellent account of the prevalence of such seemingly normal individuals who may harbor antisocial personality traits in her book *The Sociopath Next Door* (Stout, 2006). Ironically, in 1995, the author of this text lived next door to Christian and Mary Jane Longo, and socialized occasionally with the couple and was unaware of the personality traits his neighbor harbored.

BOX 12.1   Is Criminal Behavior a Genetic Disorder?

Although **criminal behavior** is not a psychiatric disorder classification, in many cases it appears to be a manifestation of one or more psychological traits. Given the high co-occurrence of antisocial personality disorder and incarceration, it would seem that identifying a genetic basis for antisocial personality disorder would coincidentally produce reasonable candidate genes for criminal behavior. On the other hand, this correlation does not suggest that a combination of genes underlying antisocial personality disorder is necessary or sufficient for the expression of criminal behavior. In other words, many individuals in prison do not exhibit this particular personality disorder. Conversely, because the number of men in prison does not exceed the estimated number who exhibit antisocial personality disorder in the general population, it is obvious that in many cases, antisocial behavior is not sufficient to produce criminal behavior.

So what other genetic elements, in addition to antisocial personality disorder, might contribute to criminal behavior? In simple terms, try to imagine a "genetic recipe" for a criminal behavior phenotype. Such a recipe might begin with alleles underlying antisocial personality disorder, which create an obvious predisposition for criminal behavior. Other genetic traits to add to the recipe might include a predisposition to addiction (discussed in Chapter 11), low levels of cognitive functioning (discussed in Chapter 10), and even a Y chromosome. Each of these factors is found in disproportionately high numbers among prison inmates compared with the general population. Does this mean that in the future, genetic screening might be used to identify criminals before they commit crimes?

Although genes are certainly important ingredients for producing criminal behavior, an equally potent set of factors is the array of environmental influences that might also contribute to such behavior. For example, a male genetically predisposed to alcoholism with a low IQ and exhibiting antisocial traits might never engage in criminal acts in an environment that is not confrontational and where all of his needs are met. However, just as these genetically controlled traits might contribute to criminal acts, they might also predispose that individual to harsher environmental conditions, such as low socioeconomic status, propensity to socialize with other alcoholics, and failure to develop a strong social support system. This example shows both the effect of environment on genes and the effect of genetic predisposition on environment selection. Such gene–environment correlations were discussed in Chapter 7.

Thus, the genetic recipe for criminal behavior seems to encourage involvement in a variety of environmental conditions that create a "social recipe" for criminal behavior. Inevitably, both genes and environment are likely to contribute to criminal behavior. Genes, however, appear to have some influence over both behaviors and environmental conditions. These observations suggest that genetic profiling might have value for providing insight into the probability of future behavioral outcomes.

Genetic influences

In attempting to identify candidate genes for obsessive–compulsive personality disorder, researchers have focused on genes linked to obsessive–compulsive anxiety disorder. The *SLITRK1* and *hSERT* genes are among those considered to be likely contributors. As a starting point for researchers, three possibilities can be considered: the less severe obsessive and compulsive symptoms seen in the personality disorder might reflect different mutations of these same genes; the same mutations of these genes interacting with different environmental factors; or different mutations of these same genes interacting with different environmental factors. Furthermore, given the broad scope of genetic functions in polygenic disorders, it is likely that variation in the phenotype is a reflection of unique combinations of several alleles that all influence the trait. Finally, it is not unreasonable to suggest that obsessive–compulsive personality disorder may be influenced by genes that are not linked to obsessive–compulsive anxiety disorder. However, this assumption would be best qualified by assuming that those different genes affect a common neurotransmitter system, or that those different genes are influenced by a common set of environmental factors.

Just as the search for candidate genes for obsessive–compulsive personality disorder has drawn heavily from research on the more severe pathological form of this phenotype, research on schizophrenia-related personality disorders has focused on findings from research on schizophrenia. Possible candidate genes for this wide array of personality disorders include all of those currently identified through linkage studies conducted on schizophrenia patients, as described in Chapter 11.

In studying the genetic basis for antisocial personality disorder, one inherent problem is its persistent comorbidity with other heritable traits, most notably addiction to alcohol and other psychoactive drugs, as noted in Chapter 11. But addictive behavior clearly may also occur with no such antisocial behaviors. As with other disorders influenced by multiple genes, these phenotypic variations suggest expression of unique allele combinations. In addition, such polygenic disorders are likely to be strongly influenced by environmental factors. Although research is in the early stages, at least two candidate genes for antisocial personality disorder have been proposed.

In a study of children, Thapar and colleagues (2005) found that one particular variant of the **catechol O-methyltransferase (*COMT*) gene** was correlated with an early-onset form of antisocial behavior. This gene is located on the long arm of chromosome 22 (22q11.2) and controls production of COMT, an enzyme utilized for metabolism of the three **catecholamine** neurotransmitters: norepinephrine, epinephrine, and dopamine. Of particular interest in this context is dopamine, which is believed to underlie sensations of pleasure and reinforcement. One theory of antisocial personality disorder is that affected individuals lack the normal pleasurable feelings evoked by social reinforcement. Such a distortion of pleasure sensations would make it difficult to distinguish feelings of guilt or remorse, leading to the primary characteristics of the disorder.

A second gene of interest in antisocial personality disorder is the **solute carrier (*SLC6A3*) gene**, located on the short arm of chromosome 5 (5p15.3). The *SLC6A3* gene regulates production of presynaptic dopamine transporter pro-

TABLE 12.5 **Some Genes Currently Implicated in Personality Disorders**

GENE (LOCATION)	ASSOCIATED PROTEIN	DISORDER	NOTES
SLITKR1 (13q31)	Slit protein	Obsessive–compulsive PD	Also implicated in obsessive–compulsive anxiety disorder
hSERT (17q11.1–q12)	Serotonin transporter protein	Obsessive–compulsive PD	Also implicated in obsessive–compulsive anxiety disorder
COMT (22q11.2)	Catechol O-methyltransferase	Antisocial PD	Particularly noteworthy in early-onset antisocial PD
SLC6A3 (5p15.3)	Dopamine transporter protein	Antisocial PD	May also contribute to addictive behaviors

teins. Just as serotonin transporter protein disruption might contribute to mood disorders, the disruption of dopamine reuptake may result in a functional imbalance of this catecholamine. In the case of dopamine, however, rather than alterations in mood, the result appears to be a disturbance of normal pleasure sensations. Interestingly, *SLC6A3*, as well as other genes that influence dopamine transmission, is frequently implicated in addictive behaviors. This is noteworthy for two reasons. The first is the extensive comorbidity of drug and alcohol addiction with antisocial personality disorder. Second, just as a lack of normal pleasurable feelings may explain failure to seek social acceptance, it may also encourage use of potent psychoactive drugs as a means of stimulating an underactive neural network to achieve pleasure and reinforcement. Table 12.5 lists and summarizes information for some of the genes currently implicated in the development of personality disorders.

Heritability

Assessing the heritability of personality disorders is somewhat difficult, both because of the range of phenotypes for each disorder subtype and because of the existence of comorbid disorders in a large percentage of cases. However, linkage studies suggest a marked distribution pattern for many of these disorders within certain families. For example, schizophrenia-related personality disorders are often intermingled in family pedigrees (Figure 12.10). In particular, these disorders are frequently identified in pedigree studies of schizophrenia (e.g., Appels et al., 2004; Kendler et al., 1993). Such findings suggest that common genetic influences probably underlie many personality disorders in this subclass as well as the more severe psychopathology of schizophrenia. It is further possible that family histories revealing both schizophrenia-related per-

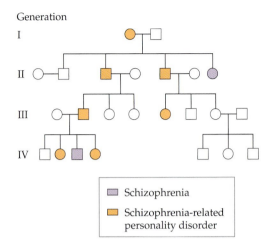

Generation

FIGURE 12.10 In this hypothetical pedigree chart, a genetically based schizophrenia-related personality disorder is present in generation I. Generations II–IV show the possible outcome of differential inheritance of alleles contributing to this psychopathology. Individuals inheriting very few contributing alleles do not meet diagnostic criteria for pathology. Individuals inheriting a moderate number of contributing alleles are diagnosed with a schizophrenia-related personality disorder. Individuals inheriting a large number of contributing alleles are diagnosed with schizophrenia. It may be assumed that offspring more severely affected than a diagnosed parent inherit some contributing alleles from their unaffected parent as well. However, it is also possible that when offspring severity is greater or the percentage of offspring affected is high (as in generations II and IV), environmental stressors shared among siblings may be responsible (e.g., family conflicts, inadequate or abusive parenting).

sonality disorders and schizophrenia represent variable inheritance of multiple contributing alleles. This theory would imply that family members who inherit fewer (or less influential) alleles would exhibit personality disorders, whereas those who inherit a greater number of (or more influential) alleles would express the schizophrenia phenotype. Similarly, such a difference in degree of symptom expression could be due to particular interactions of allele combinations or the unique environmental conditions interacting with those alleles.

Just as schizophrenia-related personality disorders appear to be associated with schizophrenia, obsessive–compulsive personality disorder appears to have a similar relationship with obsessive–compulsive anxiety disorder. More specifically, compared with unaffected control subjects, individuals with obsessive–compulsive personality disorder have a greater number of relatives diagnosed with obsessive–compulsive anxiety disorder (Nestadt et al., 2000). In addition, the inheritance pattern of obsessive–compulsive personality disorder is reflective of a polygenic disorder, occurring in pedigrees in a pattern not characteristic of single-gene inheritance.

Antisocial personality disorder arguably exhibits the most unique and complex inheritance pattern of the personality disorders. It is unique in that, despite its close association with drug addiction and alcoholism, this personality disorder does not appear to be a subtle expression of a more profound phenotype. It is somewhat complex in that researchers have identified at least two subtypes of antisocial behavior that show different inheritance patterns. More specifically, individuals tend to exhibit antisocial behaviors in either a predominantly aggressive or a predominantly delinquent (nonaggressive) fashion. In a study of early development of antisocial behavior, Eley et al. (2003) found a greater genetic influence on the aggressive form than on the delinquent form. Interestingly, these researchers also concluded that the delinquent form was greatly influenced by shared environmental factors, a finding that is rare in studies of psychopathology. One possible explanation for the relatively high degree of shared environment influence is the evocative nature of this phenotype. **Evocative traits** are those that evoke a response from the environment that in turn fosters or enhances further expression of the trait. In this case, delinquent actions frequently result in negative social consequences that, in turn, evoke future antisocial actions in an active correlation pattern (as described in Chapter 7). This particularly predictable sequence could, by its nature, create similar shared environmental features for many individuals expressing this personality disorder. Furthermore, it is likely that this form of shared environment would be closely associated with further expression of the shared trait (Figure 12.11).

FIGURE 12.11 An evocative trait evokes responses from the environment that encourage further expression of the trait. Antisocial personality disorder frequently leads to conflicts with the law. Society's response to antisocial acts is often imprisonment in a hostile environment where further expression of antisocial behavior may benefit the individual.

Genetic contributions to personality disorders are probably similar to those seen in more severe, related psychopathologies. For example, the *SLITRK1* and the *hSERT* genes implicated in obsessive–compulsive anxiety disorder have been linked to obsessive–compulsive personality disorder. Similarly, genetic anomalies associated with schizophrenia are currently being investigated as markers for schizophrenia-related personality disorders, which share features with that psychopathology.

Research on antisocial personality disorder has been influenced by genetic studies on addictive behavior. Because comorbid expression of these two traits is common, shared genetic polymorphisms are possible. In particular, the *SLC6A3* gene, which regulates dopamine transporter proteins, has been identified as a candidate gene for both addictive behaviors and antisocial personality disorder. The *COMT* gene, which regulates catecholamines (including dopamine), has also been linked to one form of early-onset antisocial behavior. It is possible that impairment of the dopamine reward system could inhibit reinforcement for appropriate social behaviors, resulting in antisocial traits. A similar functional impairment could also prompt use of psychoactive drugs to stimulate a chronically underactive dopamine reward system.

Assessing the heritability of personality disorders is difficult, both because of the range of phenotypes for each disorder and because of the existence of comorbid disorders in a large percentage of cases. Schizophrenia-related personality disorders are often found to be intermingled in pedigree studies of schizophrenia. Similarly, obsessive–compulsive personality disorder is more frequent in relatives of individuals diagnosed with obsessive–compulsive anxiety disorder than in unrelated individuals. Antisocial behavior can be subdivided into two subcategories, each with different inheritance patterns. More specifically, there is a greater genetic influence on the aggressive subtype than on the delinquent subtype.

Review Questions and Exercises

- List at least three possible reasons why more women than men report both mood and anxiety disorders.

- Describe the correlations between mood disorders and education, health, and income. Briefly describe the different ways in which these correlations could be interpreted.

- Name the mood disorder for which the *FAT* gene is considered a candidate gene. Briefly summarize the evidence suggesting that a *FAT* gene polymorphism may be involved in this particular disorder.

- Compare and contrast research findings on the prevalence of mood disorders and of anxiety disorders.

- Briefly explain why the most common treatment for anxiety disorders has not led to a fruitful starting point for behavior genetics research.

- Discuss the evidence suggesting that some personality disorders and some more severe psychopathologies have shared genetic influences.

- List several genetic factors that could be considered precursors to a theoretical trait of "criminal behavior." Briefly explain the contribution of each factor.

- Describe the features of antisocial personality disorder that make it unique among the personality disorders discussed in this chapter.

PART IV
OTHER GENETIC INFLUENCES, COUNSELING, AND THE FUTURE

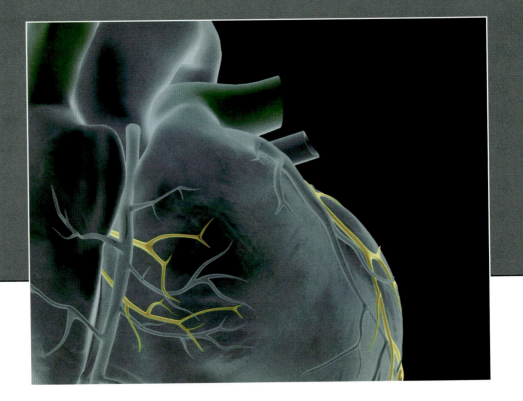

13

Beyond Psychopathology: General Health Risks and Sex Chromosome Abnormalities

Up to this point, the focus of this text has been on behaviors directly associated with cognitive function, mood, and personality. Dysfunction of these behaviors can be directly observed in psychopathologies. Beyond these psychopathologies, some behavioral changes may also be evaluated in the context of their relationship to physical disorders. For example, disordered eating behaviors such as anorexia and bulimia were presented as psychopathologies in Chapter 11. At the other end of the behavioral spectrum, however, is excessive eating that may be associated with obesity. Although obesity is not typically considered a psychologically based disorder, there are, in some cases, interesting psychologically influenced behavioral contributions to that disorder. Similarly, hypertension and other cardiovascular diseases are certainly affected by genetic influences on peripheral system functions, but psychological traits may also contribute to the development of these physical illnesses. In fact, many illnesses once considered to be purely based in peripheral physiology are now known to be influenced by psychological factors. This chapter also describes how sex chromosome inheritance patterns differ from inheritance patterns described for autosomes. Included among these differences is the addition or deletion of chromosomes, a phenomeon that occurs much more frequently in sex chromosomes than in autosomes, and with far less severe consequences. More specifically, sex chromosome trisomy and monosomy produce notable physical effects with relatively mild cognitive impairments. However, because

To many readers, it may be a somewhat novel concept that changes in immune system function can affect neural processes and, ultimately, behaviors associated with mood and emotions. But consider for a moment the inverse relationship between these systems—that is, the effects of the nervous system on immune system function. One example is the ability of psychological stress to impair immune system activity. Stress is a common contributing factor to many disorders that are caused or exacerbated by compromised immune function. Such ailments include viral illnesses, dermatological conditions, and dysfunction of the digestive, respiratory, and cardiovascular systems. The primary reason for this casual relationship between psychological stress and physical illness is that neural responses to stress include alterations in endocrine system activity, including increased release of the glucocorticoid hormone cortisol. Recall from Chapter 12 that chronic elevation of cortisol has been implicated as a precipitating factor in several mood and anxiety disorders. In addition, elevated cortisol concentrations are known to impair immune function. Scapagnini (1992) notes that most organismal functions are influenced by one or more of the following three systems: the nervous system, the endocrine system, and the immune system. Furthermore, he suggests that the highly interactive relationship of these systems warrants an integrative approach, sometimes referred to as psychoneuroendocrinoimmunology (Box 13.1).

Just as psychopathology has been linked to some forms of physical illness, psychological functions may have the potential to improve physical health. As noted, both anxiety and depression appear to be correlated with impaired immune function. Conversely, meditation and positive imagery have the potential to improve immune responses in some individuals (see Gruzelier, 2002 for review). It has long been accepted that such relaxation exercises may be used to down-regulate "involuntary" peripheral functions such as blood pressure, heart rate, and oxygen consumption. It appears that reducing these metabolic functions may conserve energy that can then be redirected to the immune system. In this regard, relaxation techniques appear to have a functional physiological end result similar to the depressive state induced by proinflammatory cytokines. Gruzelier (2002) also takes note of research suggesting that relaxation techniques can directly lower adrenal release of stress hormones. This finding further suggests that common genetic influences underlie some behaviors as well as physiological functions associated with general health.

Psychophysiological illness

The term **psychosomatic disorder** was once commonly used to describe a physical illness brought on, or exacerbated, by psychological factors such as anxiety or depression. Unfortunately, this term was frequently misused and misinterpreted as a synonym for **hypochondria**, which is a state of imagined illness

BOX 13.1   Psychoneuroendocrinoimmunology

The term **psychoneuroendocrinoimmunology (PNEI)**, though cumbersome, aptly describes the study of interactions between psychology, the nervous system, the endocrine system, and the immune system. PNEI certainly ranks among the more recently developed fields of research, but its origins are based on four long-established areas of study. In fact, the conceptualization of PNEI as an independent field of research does not reflect a profound revelation brought forth by new discoveries. Instead, it represents a predictable progression in attempts to organize the existing data that point to closely linked and interrelated functions of the brain and the periphery.

This movement toward a more unified approach to studying the basis for normal behaviors, as well as pathologies, parallels changes in the field of genetics. Think back for a moment to the discoveries of the Human Genome Project described in Chapter 4. Prior to the Human Genome Project, estimates of the number of human genes typically exceeded 100,000. As data accumulated, that number was cut in half, and then cut in half again, eventually bringing us to the current estimate of 20,000–25,000 genes. One revelation based on the discovery of this relatively small number of genes is that individual genes must control a much broader array of functions than was originally believed. In addition, researchers have developed a greater appreciation of the potential for interactions between genes in controlling behaviors. Future discoveries of genes that influence multiple systems, and of gene interactions that mediate complex behaviors, are likely to encourage more multifaceted approaches to explaining behavior genetics such as that adopted by PNEI.

that has no physical basis. The term **psychophysiological illness** is now frequently used to describe an illness in which the physiological symptoms are influenced by mental processes. In the case of psychophysiological illness, medical examinations often find no direct physical or organic cause. Instead, symptoms typically appear when the patient becomes angry, depressed, anxious, or experiences a high degree of guilt.

Interestingly, in psychophysiological illnesses, the psychological state and resulting physical manifestations often vary among individuals. For instance, hypertension that results from feelings of anger in one individual may result from feelings of anxiety in another. Likewise, the same levels of anxiety that evoke hypertension in one individual may cause an asthma attack in another. Such variation suggests that the interplay between psychological and physical functions is largely dependent on the vulnerabilities of the individual. For example, a genetic predisposition to respiratory distress could make an asthmatic response to stress hormone elevation most likely, whereas the same stress hormone elevation would produce hypertension in a person predisposed to vascular distress. Psychophysiological illnesses may be acute (migraine headache, asthma attack, skin rash) or chronic (ulcer, hypertension, skin rash).

FIGURE 13.2 Health psychology considers the interplay between physical fitness and mental health. By-products of physical illness, such as elevated cytokines, have the potential to trigger or exacerbate psychopathologies. Similarly, by-products of psychopathologies, such as elevated stress hormones, can contribute to physical illness. In this regard, genes that directly contribute to physical illness may predispose individuals to some psychopathologies, and vice versa. This intimate interrelationship further underscores the need for genetic research if we are to fully understand the basis for both psychological and physical maladies.

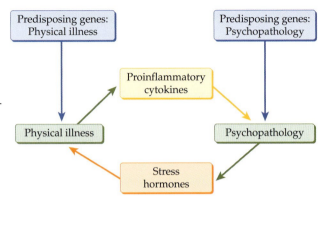

With this understanding that certain psychopathologies can influence the onset of physical illness, and that physical illness has the potential to affect psychopathology in similar ways, a closely intermingled relationship becomes apparent (Figure 13.2). The genesis of many such interrelated physical and psychological functions lies in genetic predispositions. As researchers learn more about genetic contributions to both physical and psychological maladies, increased understanding of a large number of related disorders is likely to follow. Thus, the range of therapeutic approaches may accelerate rapidly with future discoveries in the field of behavior genetics.

Cardiovascular Disease

As noted above, several health risks are directly influenced by psychological factors. Psychophysiological illnesses with potentially lethal consequences include ulcers, asthma, and cardiovascular diseases. For the purposes of this text, cardiovascular diseases are used as primary examples of psychophysiological illnesses. Keep in mind, however, that the principles discovered through research into these diseases may be readily applied to other illnesses as well.

Hypertension

The most common form of chronic cardiovascular disease is **hypertension**, characterized by persistent elevated blood pressure within the arteries. Although it is categorized as a cardiovascular disease, hypertension affects an array of other organs, including the kidneys, eyes, and brain. In addition, it is frequently a precursor to some forms of acute cardiovascular disease.

Compared with the general population, a greater percentage of patients diagnosed with either depression or anxiety disorders are classified as clinically hypertensive (Löwe et al., 2004). Johannessen (2006) recently reported similar findings of elevated rates of hypertension in patients with either anxiety disorders or bipolar disorder. Johannessen's study used patients diagnosed with schizophrenia as a control group. The results showed that schizophrenia was not associated with hypertension, indicating a relatively specific relationship between this cardiovascular disease and certain psychopathologies.

Despite the correlations between some psychological and physical maladies that have been established, there is little evidence to show a causal relationship. In other words, it is uncertain whether hypertension contributes to the development of psychopathologies, psychopathologies contribute to the development of hypertension, or some shared underlying factor(s) contributes to both.

Acute cardiovascular disease

The three most common types of acute cardiovascular disease are coronary heart disease (which includes heart attack and angina pectoris or chest pain), stroke, and heart failure. Although it may seem reasonable to assume that anxiety disorders could result in acute coronary dysfunction, there is little evidence to indicate a direct causal relationship. In particular, panic attacks and panic disorder are characterized by symptoms reminiscent of coronary heart disease. However, the risk of heart attack does not appear to be increased as a result of acute anxiety, such as that experienced with panic disorder. In fact, no form of psychopathology has been directly linked to an increased risk of acute coronary disease. However, as noted in the previous section, hypertension is known to increase the risk of acute cardiovascular disease. In particular, chronic hypertension dramatically elevates the risk of stroke and some forms of coronary heart disease. In this regard, the direct correlation of hypertension with certain psychopathologies establishes an indirect correlation between those psychopathologies and elevated susceptibility to acute cardiovascular disease.

Influence of personality type

There was once a notable emphasis placed on the contributions of personality type to the development of both chronic and acute cardiovascular disease. The term **type A** was often used to describe a personality marked by aggression, impatience, a sense of time urgency, and a desire to achieve recognition and advancement. A variety of self-administered tests can be used to establish type A personality as an enduring characteristic. When this personality type was first

FIGURE 13.7 Discrepancies between the X and Y chromosomes influence gene inheritance patterns. (A) Paired autosomes, such as chromosome 12, depicted here, share homologous regions of DNA. Pairing of these genes allows alleles to be expressed as dominant, recessive, or in partial dominant fashion. (B) Paired X chromosomes share this same feature. (C) In males, however, the pairing of the structurally dissimilar X and Y chromosomes creates unique patterns of gene expression.

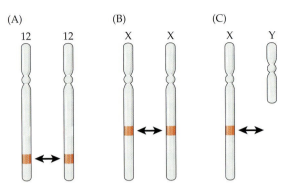

gous gene to influence the phenotype. When a gene is inherited on the X chromosome, on the other hand, a homologous gene is present in females (XX), but not in males (XY) (Figure 13.7). These unique discrepancies create interesting inheritance patterns as well as sex-specific phenotypes in some cases.

X-LINKED RECESSIVE   Recessive dysfunctional alleles on the X chromosome affect males more frequently than females. In females, the unaffected allele on the homologous chromosome typically compensates for the dysfunctional allele. Because males lack such an unaffected allele on the Y chromosome, they express the dysfunctional recessive phenotype. Only females who inherit the recessive allele from both parents express the dysfunctional phenotype (Figure 13.8). In terms of inheritance patterns, a male cannot pass an X-linked recessive disorder on to his male offspring. However, all female offspring of an affected male will be carriers because they will inherit one copy of the dysfunctional allele. This means that female offspring of an affected father have a 50% chance of expressing the phenotype if their mother is also a carrier of the allele. A woman who carries an X-linked recessive disorder also has a 50% chance of having sons who are affected.

Perhaps the most frequently cited examples of X-linked recessive disorders are some forms of color blindness (Figure 13.9). The most common form of **hemophilia**, a disease characterized by a lack of blood clotting, is also an X-linked recessive disorder. Similarly, the most common form of muscular dystrophy is caused by a recessive allele on the X chromosome. **Duchenne muscular dystrophy** is the most prevalent in a family of dystrophy diseases characterized by progressive muscle weakness leading to chronic degeneration of muscle fibers.

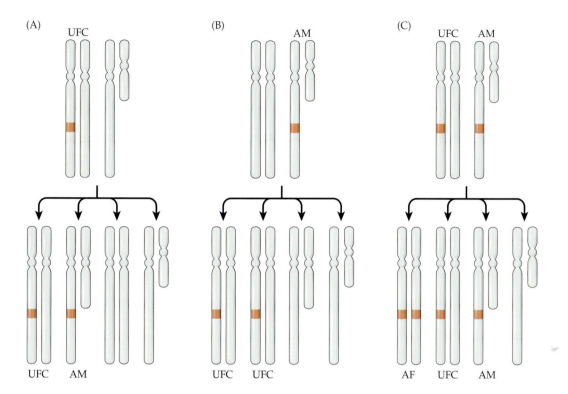

FIGURE 13.8 Inheritance patterns and expression of X-linked recessive genes. (A) 50% of the female offspring of an unaffected female carrier (UFC) will also be unaffected female carriers, and 50% of her male offspring will be affected males (AM). (B) 100% of the female offspring of an affected male will be unaffected female carriers. (C) 50% of the female offspring of an affected male and an unaffected female carrier will be unaffected female carriers, and 50% will be affected females (AF). 50% of the male offspring of these parents will be affected males.

Fragile X syndrome (described in Chapter 10) is caused by a triplet repeat expansion on the X chromosome. Although fragile X syndrome is not considered a typical example of an X-linked recessive disorder, it has been shown that as many as one-third of cases follow a pattern of inheritance indicative of recessive allele expression (Fryns, 1984). The unique features of the fragile X mutation probably account for some of the cases not predicted by a normal X-linked recessive inheritance pattern. For example, this particular allele sometimes exhibits premutation, characterized by increasing expansion of the triplet repeat and increasing expression of the fragile X phenotype over several generations.

FIGURE 13.9 In females, if one X chromosome carries an abnormal red–green color receptor gene, an unaffected gene on the other X chromosome can direct development of normal color vision. A similarly dysfunctional gene in males, however, has no homologous gene to direct normal red–green color receptor formation. In this case, red–green color blindness can develop. Individuals with this type of color blindness see the number "21" in this figure, whereas a person with normal color vision sees the number "74."

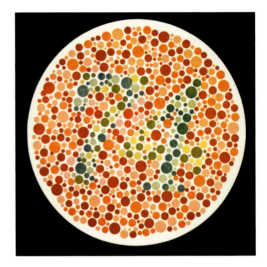

X-LINKED DOMINANT   Very few disorders have an X-linked dominant inheritance pattern. As with X-linked recessive disorders, a man with an X-linked dominant disorder cannot pass it on to his male offspring. However, all female offspring of a man with an X-linked dominant disorder will express the phenotype. Offspring (male and female) of a woman with an X-linked dominant disorder each have a 50% chance of being affected. The prevalence of X-linked dominant disorders is greater in females than in males because females inherit two X chromosomes. In addition, some X-linked dominant disorders are lethal in males, but not in females, who may partially compensate for the dysfunctional allele with a normal allele on the homologous chromosome. The fact that males are less likely to survive with X-linked dominant disorders also contributes to the high female-to-male ratio of affected individuals.

Hypophosphatemia, a condition characterized by low levels of phosphate in the blood, can be caused by alcohol abuse or by poor phosphate absorption from dietary intake. One form of hypophosphatemia caused by kidney dysfunction, however, is inherited as a dominant X-linked disorder.

Aicardi syndrome, most notably characterized by partial or complete absence of the corpus callosum, is also inherited as an X-linked dominant disorder. The corpus callosum is the primary structure that transfers information between the brain's two hemispheres. Other symptoms of Aicardi syndrome may include microcephaly (a small brain), cerebral ventricle enlargement, and porencephalic cysts (large gaps in the brain tissue) (Figure 13.10). Aicardi syndrome is fatal in males. Females born with this syndrome typically suffer from seizures and developmental delays that culminate in moderate to severe mental retardation.

(A)

Corpus callosum

(B)

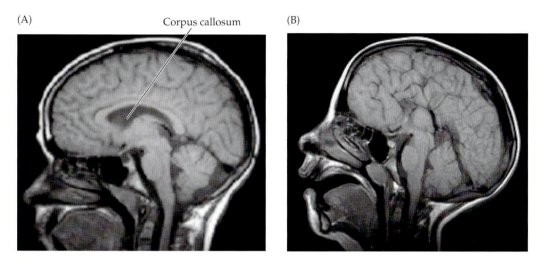

FIGURE 13.10 Aicardi syndrome is caused by an X-linked dominant allele. (A) Brain MRI showing a normally developed corpus callosum. (B) MRI showing a brain lacking a corpus callosum. This type of abnormal development is a defining characteristic of Aicardi syndrome.

Y-LINKED Few Y-linked disorders have been identified. Because they are caused by mutations of genes on the Y chromosome, only males are affected. In addition, all male offspring of an affected father inherit these disorders. Although one form of hearing impairment may be caused by a mutation on the Y chromosome (Wang et al., 2004), no disorders with complex behavioral phenotypes have been traced to this chromosome. Instead, because the vast majority of genes on the Y chromosome control spermatogenesis and testicular function, most Y-linked mutations are associated with impaired fertility. As such, these mutations are rarely transmitted to offspring.

Deletions and additions

The concepts of monosomy and trisomy were presented in Chapter 10. In particular, it was noted that lack of a chromosome from any autosome pair is lethal, usually during fetal development or shortly after birth. An extra autosome is also typically lethal; the most notable exception is Down syndrome, caused by the addition of an extra chromosome 21. Deletions or additions of sex chromosomes, however, are less frequently lethal. The exception to this rule is the absence of an X chromosome in the presence of a Y chromosome. There have been no reported cases of a fetus developing with Y chromosome monosomy. The reason for this is probably the small amount of genetic material on the Y chromosome, as described

FIGURE 13.11 Features typical of the X0 monosomy genotype, commonly known as Turner syndrome.

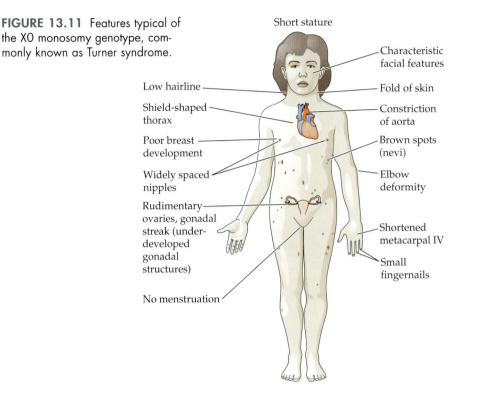

Short stature

Characteristic facial features

Low hairline

Fold of skin

Shield-shaped thorax

Constriction of aorta

Poor breast development

Brown spots (nevi)

Widely spaced nipples

Elbow deformity

Rudimentary ovaries, gonadal streak (underdeveloped gonadal structures)

Shortened metacarpal IV

Small fingernails

No menstruation

above. It appears that the X chromosome contains vital genetic information that is responsible not only for sexual development, but also for survival of the fetus to birth. The absence of that information, like the absence of an autosome, is lethal.

TURNER SYNDROME   It has been estimated that about 1 in every 2,000 females is born without a Y chromosome (X0 genotype) (see Gravholt, 2008). Although the absence of an X chromosome is lethal, the absence of a Y chromosome is far less consequential. Development of an X0 individual results in **Turner syndrome**, which is characterized by relatively normal external female features. However, lack of a second sex chromosome does cause several notable abnormalities, including short stature and ovarian abnormalities (typically resulting in infertility). Elbow, hand, and finger malformations may also occur (Figure 13.11). In some cases, individuals with Turner syndrome show developmental delays, and IQ scores below the normal range are not uncommon.

The features of Turner syndrome suggest that a single X chromosome carries enough genetic information for nearly normal female development. It also indicates that the genetic information carried on the Y chromosome, though es-

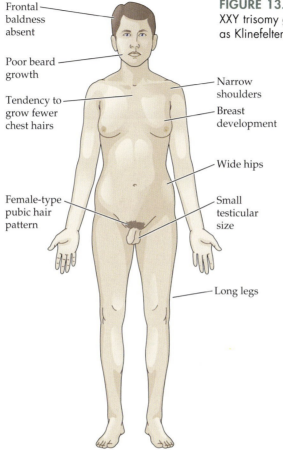

Frontal baldness absent

Poor beard growth

Tendency to grow fewer chest hairs

Female-type pubic hair pattern

Narrow shoulders

Breast development

Wide hips

Small testicular size

Long legs

FIGURE 13.12 Features typical of the XXY trisomy genotype, commonly known as Klinefelter syndrome.

sential for development of male characteristics, is not necessary for survival. Considering the small amount of genetic information carried on the Y chromosome, the redundancy of that information, and the specificity of that information to spermatogenesis and testicular function, this result is not surprising.

KLINEFELTER SYNDROME   **Klinefelter syndrome** (XXY genotype) is the most common abnormality of the sex chromosomes, affecting approximately 1 in every 650 men (Bojesen et al., 2003). The addition of an X chromosome to the normal male karyotype results in impeded male development with some female attributes. Physical features include male genitalia that are typically undersized, sparse facial hair, female-typical fat distribution, and some breast tissue development (Figure 13.12). Affected individuals usually have low sperm production, resulting in infertility. Although mental retardation is not considered a primary feature of Klinefelter syndrome, developmental delays and cognitive impairments are common and may culminate in below-average IQ scores.

The features of Klinefelter syndrome indicate that the presence of a Y chromosome is necessary and sufficient to produce male development when it is paired with at least one X chromosome. It appears that male development is impeded and some feminization occurs in affected individuals as a result of excess expression of X-linked genes. It is also noteworthy that though Klinefelter syndrome may be characterized by infertility and reduced libido (Nicholls and Anderson, 1982), there is no indication that this condition increases the likelihood of homosexual behaviors in affected individuals.

XYY SYNDROME   The prevalence of the XYY genotype is generally estimated to be about 1 in 1,000 males. Because the features associated with the XYY genotype are less profound than those of other trisomies, some researchers have argued against using the term "syndrome." In fact, the most prominent physical feature of the XYY phenotype is an increase in height relative to other family members. Although some reports have cited a greater incidence and severity of acne (presumably resulting from increased androgen levels), this physical feature is not apparent in many XYY individuals. In fact, androgen levels are typically unaffected by the addition of a Y chromosome, and fertility rates in these individuals are usually normal. In terms of cognitive development and intellectual capacity, there is some suggestion that the addition of a Y chromosome is correlated with a higher rate of learning difficulties, particularly in language development. As a result, below-average IQ scores are more common in XYY than in XY individuals. There is also some evidence that XYY syndrome is associated with a higher frequency of criminal behavior. However, it is unclear what particular XYY associated trait might directly contribute to criminal behavior (Box 13.2).

The lack of profound physical alterations and cognitive effects associated with an additional Y chromosome is not surprising. Given that much of this chromosome controls spermatogenesis and testicular function, it is reasonable to conclude that redundancy in this information would have little effect on cognitive development. In addition, the same palindromic organization of base sequences that allows functional recombination within a single Y chromosome may also be conducive to normal recombination when it is paired with a second Y chromosome. Such capability would further explain the lack of negative effects on reproductive function in XYY individuals. Finally, the increased height of affected individuals suggests that production of growth hormone may be influenced by genes within the male-specific region of the Y chromosome.

BOX 13.2   Is XYY Syndrome Associated with Criminal Behavior?

Is the addition of a second Y chromosome linked to an increase in the likelihood of criminal behavior? Several case studies in the late 1960s suggested that an extra Y chromosome might be linked to expression of antisocial behavior. These reports led to a more extensive series of survey studies designed to test the hypothesis that XYY males were more inclined toward criminal behavior than were XY males. Interestingly, this hypothesis was readily supported, with findings from several research groups showing a significantly higher percentage of XYY males in prison populations than in samples of nonincarcerated men (see Witkin et al., 1976 for review).

There is an all-too-easy tendency to interpret these data as suggesting that, just as a single Y chromosome is associated with androgen production and thus with increased potential for aggression, the presence of a second Y chromosome may elevate aggressive potential to a point of evoking criminal acts. To test this logical hypothesis, Witkin and colleagues assessed a sample of XYY men to determine, first, whether their rate of criminal activity exceeded not only that of XY males, but also that of XXY males, and second, whether their criminal acts were associated with increased aggression.

The findings were extremely enlightening. First, although Witkin and colleagues did replicate the earlier findings by showing a significantly higher rate of criminal convictions in a sample of XYY males (41.7%) than in XY males (9.3%), they also found an elevated rate of criminal convictions among XXY males (18.8%). In fact, in their report, the difference in criminal conviction rate between the XYY and XXY samples was not statistically significant. Perhaps even more interesting, the percentage of crimes involving aggression toward others did not differ significantly between any of the three samples of men.

So what are the implications of these findings? One possible explanation is that the increased rate of criminal convictions in XYY males is more closely linked to the cognitive deficits associated with a sex chromosome trisomy than with increased aggression. Decreased cognitive ability could contribute to both a propensity toward delinquent behavior and a decreased ability to elude law enforcement personnel. This hypothesis would also help explain why XYY males do not differ significantly from XXY males in their levels of criminal activity.

BEYOND THE COMMON VARIANTS   It should be noted that sex chromosome abnormalities beyond X0, XXY, and XYY do exist. It is generally estimated that about 1 in 1,000 females is born with **trisomy X** (XXX). These women are somewhat taller than XX females and have some minor characteristic facial features. They are infertile and prone to developmental delays, and in some cases suffer from seizures. Although their IQ may be below average, few are categorized as mentally retarded.

Although it is far less common, the addition of two or more sex chromosomes can occur without lethal consequences. **XXYY syndrome** is often cited as a variation of Klinefelter syndrome. It is likely that fewer than 1 in 15,000 males is born with this genotype. Although this syndrome does share some characteristic features with Klinefelter syndrome, affected individuals tend to be taller (probably resulting from influences of the additional Y chromosome) and exhibit more dermatological disorders, such as skin ulcerations. Developmental delays and cognitive impairments are more apparent in this disorder than in Turner syndrome or sex chromosome trisomies. These more severe disabilities probably result from the addition of excessive genetic instructions that interfere with normal levels of gene functioning.

Other rare syndromes include **XXXX**, **XXXYY**, and **XXXXY** syndromes. These syndromes result in phenotypes that appear as further exaggerations of the symptoms of trisomy X and Klinefelter syndrome. More important than the specific characteristics of these syndromes is the understanding that even such extensive repetition of these genes is not always lethal. This feature is unique to the sex chromosomes, as addition of two extra autosomes is always lethal. Although the sex chromosomes have the obvious essential function of determining and controlling the sex of the individual, their contributions to more vital features of development appear to be fairly limited.

Chapter Summary

One way in which psychological distress can directly impair physical health is through immune function suppression associated with chronic anxiety and depression. Psychopathologies may also have indirect effects that increase vulnerability to illness. For example, both depression and anxiety are characterized by insomnia, abnormal eating habits, and increased risk of drug or alcohol abuse. Each of these characteristics has the potential to adversely affect general physical health.

There is also evidence that some peripheral illnesses can contribute to psychopathologies. Acute increases in proinflammatory cytokines can, for example, disrupt neuronal signaling and, consequently, psychological function. Chronic inflammatory disorders, which are characterized by elevated cytokine concentrations, are often comorbid with major depression. Psychoneuroendocrinoimmunology is the study of the complex interactions between psychological function, neural activity, endocrine system responses, and the immune system. Research from this and related fields has shown that just as impaired psychological functioning can contribute to illness, enhanced psychological functioning can reduce illness. Relaxation techniques and positive imagery are among the psychological techniques used to improve physical health. These techniques appear to work by directly altering levels of adrenal hormones, leading to improved immune function.

The term psychophysiological illness is used for illnesses in which physical symptoms result primarily from emotional states such as anger, depression, anxiety, or guilt. Psychophysiological illnesses may be acute or chronic. Interestingly, from one individual to the next, similar emotional states may result in different physical manifestations (e.g., anxiety that causes an asthma attack in one individual might produce a skin rash in another).

When compared with the general population, a greater percentage of patients diagnosed with depression, anxiety disorders, or bipolar disorder are clinically hypertensive. It is uncertain, however, whether hypertension contributes to the psychopathologies, the psychopathologies contribute to hypertension, or some shared underlying factor(s) contributes to both. In any case, chronic hypertension is frequently a precursor to acute cardiovascular disease. Although anxiety disorders are characterized by symptoms reminiscent of coronary heart disease, there is probably no direct causal relationship. In fact, no psychopathology has been directly linked to an increased risk of acute cardiovascular disease. Although the incidence of coronary heart disease in individuals with type A personality is not greater than in the general population, these individuals tend to suffer heart attacks at a younger age. This may be because type A personality predisposes individuals to early and more frequent exposure to risk factors such as smoking, alcohol consumption, or strenuous physical activity.

Obesity is a well-established risk factor for a wide range of physical ailments. It is most prevalent in industrialized societies that foster sedentary behavior and excess caloric intake. The prevalence of obesity has increased dramatically over several generations, seemingly contradicting the fundamental theories of evolution. This contradiction is explained by the theory that human genes favoring fat accumulation have been historically beneficial for survival. Furthermore, though it is detrimental to physical health, obesity has limited effects on reproductive success in humans.

One recent study found a concordance rate of 82% for obesity in monozygotic twin pairs and a rate of 46% in dizygotic twins, indicating a strong genetic influence. However, dozens of single-gene mutations, and hundreds of gene combinations, may contribute to an obese phenotype. Contributing genes could influence traits such as food consumption rate, food intake volume, caloric intake, physical activity, and basal metabolism. In addition, several psychopathologies are characterized by altered food intake, suggesting the possibility of an indirect effect of genes that affect mood, anxiety, and cognitive abilities.

It is widely believed that the X and Y chromosomes arose from an autosome pair closely resembling the X chromosome. According to this theory, some regions of the Y chromosome lost the ability to undergo recombination with homologous regions on the X chromosome, leading to the diminished size of the

Y chromosome. In fact, approximately 95% of Y chromosome DNA, sometimes referred to as the male-specific region, is incapable of recombination with the X chromosome. Interestingly, many base sequences throughout this male-specific region are duplicated in nearly identical reverse order. It appears that these palindromic sequences allow recombination and repair to occur within the Y chromosome. Since genes within the male-specific region primarily influence male fertility, these capabilities are essential for the continued viability of this chromosome and for the persistence of sexually reproducing species.

Recessive alleles on the X chromosome affect males more frequently than females. X-linked recessive disorders include some forms of color blindness, one form of hemophilia, and Duchenne muscular dystrophy. In addition, about one-third of fragile X cases follow the pattern of an X-linked recessive disorder. X-linked dominant disorders are rare and are frequently lethal in males. They include hypophosphatemia (low levels of phosphate in the blood) and Aicardi syndrome (partial or complete absence of the corpus callosum). One of the very few Y-linked disorders is a form of hearing impairment. No disorders with complex behavioral phenotypes have been traced to the Y chromosome, probably because it is primarily responsible for spermatogenesis and sperm motility.

About 1 in every 2,000 females is born with Turner syndrome (X0), which is characterized by short stature, ovarian abnormalities, and some developmental delays in an otherwise relatively normal female phenotype. Approximately 1 in every 650 males is born with an additional X chromosome, resulting in Klinefelter syndrome, characterized by undersized male genitalia, sparse facial hair, female-typical fat distribution, and some breast tissue development. Developmental delays and cognitive impairments are also common in this syndrome. About 1 in 1,000 males is believed to be born with an XYY genotype, characterized primarily by an increase in height relative to other family members. Some XYY individuals also have below-average IQ scores, most likely associated with language skill deficiencies. Trisomy X (XXX) is believed to affect about 1 in 1,000 females and results in infertility, below-average IQ scores, and in some cases, seizures. Other sex chromosome abnormalities (XXYY, XXXX, XXYYY, XXXY) affect only a small number of individuals and are characterized by features similar to those seen in either Klinefelter or XXX syndrome.

Review Questions and Exercises

- Explain how psychopathologies such as depression and anxiety can directly contribute to physical illnesses. Give several examples of traits associated with these same psychopathologies that might indirectly contribute to physical illnesses.

- Just as psychological dysfunction can impair physical health, physical ailments can contribute to some forms of psychopathology. Give an example of one such physical ailment and describe how it may affect psychological function.

- Explain briefly why an anxiety disorder is more likely to contribute indirectly to coronary heart disease than it is to directly cause coronary heart disease.

- Explain briefly why obesity is so common in some cultures, particularly given that its effects on health do not seem compatible with evolutionary theory.

- List at least five different traits that could contribute to obesity. List at least three different psychopathologies that have altered food intake as a primary characteristic.

- The sex chromosome pairing is the only one that does not involve chromosomes of equal size and with equivalent genes. Explain how this chromosome pairing might have developed.

- Summarize inheritance patterns, showing transmission from both parents to one male and one female offspring, for each of the following: (1) Y-linked dominant, (2) X-linked dominant, (3) Y-linked recessive, (4) X-linked recessive.

- Briefly characterize each of the following in terms of sex chromosome abnormalities and characteristic features: (1) Turner syndrome, (2) Klinefelter syndrome, (3) XXX syndrome.

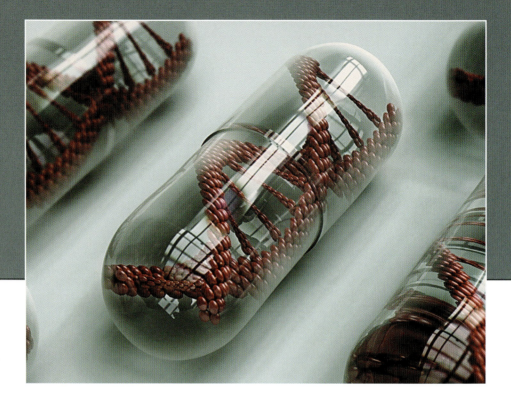

14

Genetic Counseling, Applied Pharmacogenomics, and Gene Therapy

U p to this point, this text has presented a vast array of information, ranging from basic descriptions of DNA to discussions of complex theories about how multiple genes may interact with one another and with the environment to produce particular phenotypic variations. Although the topics of the previous 13 chapters have been far reaching, one topic that has been conspicuously absent is a detailed assessment of how current knowledge is being used to counsel and treat individuals affected by genetic disorders. Elucidating the current state of genetic counseling and gene-based therapy is important for several reasons. First, such applications represent a primary goal of basic research in the field of behavior genetics. Second, the present state of genetic counseling and related therapies offers guideposts for future basic research by highlighting particular areas in need of greater clarity. Third, beyond their obvious connection to basic research, genetic counseling and gene-based therapies are growing fields, deserving further explanation in their own right. Genetic counseling, for example, is a rapidly progressing and expanding profession. In just the past 20 to 30 years, the volume and complexity of the information available for genetic counseling has expanded at a remarkable pace. The history of gene-based therapies begins even more recently, making discussions of the early and encouraging findings in this area of research tentative but exciting.

Genetic Counseling

Tracing the history of any profession to its origin is a speculative enterprise at best, in part because of the difficulty of identifying the defining moment when an independent field is officially recognized. Clearly, informal genetic counseling had been undertaken since information first came to light about hereditary transmission of certain single-gene disorders. Such counseling, however, was delegated to general physicians and psychiatrists with no particular training or specialization in genetics, and it was typically limited to patients at risk for single-gene disorders with Mendelian patterns of inheritance that were well established and relatively simple to explain. As the benefits of counseling individuals on the nature of hereditary diseases became clear, some physicians, psychiatrists, and counselors began to specialize in that area. This movement advanced gradually over the years, and it is impossible to be certain when the first person assumed an exclusive role as a genetic counselor. It has been established, however, that by the early 1970s, several genetic researchers and counselors had begun discussions about the need for a professional society to formally recognize the emerging field of genetic counseling (Heimler, 1997).

In response to those initial discussions, the **National Society of Genetic Counselors** (**NSGC**) was founded in 1979. In 1981, the **American Board of Medical Genetics** (**ABMG**) was established and became the sanctioning body for certification of genetic counselors. Over the next decade, the accumulation of genetic counseling research inspired the development of a scientific journal dedicated to this field. The first volume of the *Journal of Genetic Counseling* (Springer, Netherlands) was subsequently published in 1992, and the journal continues to serve as a primary source for communicating findings in the field. In 1993, the rapidly expanding ABMG was restructured. To accommodate growing interest in the field, members of the ABMG voted to establish the **American Board of Genetic Counseling** (**ABGC**), which now certifies genetic counselors and accredits genetic counseling programs. This historical timeline reflects just how recently the field of genetic counseling was formalized and how rapidly it has grown.

What is genetic counseling?

Although it might seem that a definition of genetic counseling should be intuitive, consider for a moment the complexity of this profession. Inherent among the responsibilities of a genetic counselor is, at the very least, the ability to convey information about genetics, heredity, the normal range of phenotypes, the probability of disease inheritance, and the management and treatment of genetic diseases. It is apparent that this profession does not lend itself to a concise definition. In fact, the NSGC recently formed a task force charged with drafting a formal definition of genetic counseling (Resta et al., 2006). The Genetic Counseling Definition Task Force put forth the following description, which was subsequently adopted by the NSGC:

> Genetic counseling is the process of helping people understand and adapt to the medical, psychological and familial implications of genetic contributions to disease. This process integrates the following:
>
> • Interpretation of family and medical histories to assess the chance of disease occurrence or recurrence.
>
> • Education about inheritance, testing, management, prevention, resources and research.
>
> • Counseling to promote informed choices and adaptation to the risk or condition.

This definition nicely embodies the scope of the primary responsibilities of genetic counselors. It also helps individuals who are seeking genetic counseling to understand the range, as well as the limitations, of what they might expect from these professionals.

What kinds of information do genetic counselors interpret?

It is apparent that genetic counselors assume a wide range of responsibilities with their job title. But what is the source of information for the services they provide? Clearly, the first 13 chapters of this text could be considered an overview of the staple knowledge required to enter the field. But such knowledge constitutes only a general information base. For each individual with whom they work, genetic counselors must gather case-specific information. Just as physicians use readings of reflexive responses, vital functions, and blood test results from their patients, genetic counselors require relevant data for their evaluations.

A **family history** of physical or psychological dysfunction provides essential data for genetic counselors. Pedigree charts offer counselors an invaluable tool for mapping and interpreting family histories (Box 14.1). Because clients are often responsible for providing their own family history information, it might seem that the self-awareness brought about by this data-gathering process would reduce the need for a genetic counselor. However, in a recent review, Finn and Smoller (2006) concluded that patients with a family history of psychiatric illness often underestimate the risk of illness in their children. In contrast, both spouses and siblings of affected individuals tend to overestimate the risk of illness in their children. These findings emphasize the need for a qualified genetic counselor who can interpret family histories and provide concerned individuals with the most accurate information available regarding the risk of illness in offspring.

BOX 14.1 A Pedigree Is Worth a Thousand Words

In Chapter 8, the use of pedigree charts for mapping gene transmission was briefly described. These diagrams have obvious value for genetic counselors seeking to summarize a family history of genetic dysfunctions. As the old adage states, "A picture is worth a thousand words." Pedigree charts use symbols to identify multiple characteristics of individuals and to show, at a glance, how dozens of individuals are related. A clearly composed pedigree chart can thus provide counselors or medical practitioners with a simple pictorial summary of a complex family history of trait expression.

With this in mind, it seems obvious that symbols in pedigree charts should be written using some standardized format—and they are. However, the establishment of those standards was more recent than most people realize. In fact, numerous inconsistencies in the use of pedigree symbols, even within the scientific literature, persisted into the 1990s. In response to those inconsistencies, and to the growing use of pedigree charts by a wide range of health professionals, the National Society of Genetic Counselors (NSGC) formed a Pedigree Standardization Task Force (PSTF) with the primary charge of developing a standardized human pedigree nomenclature. The PSTF

	Male	Female	Sex Unknown
Individual	b. 1925	30 y	4 mo
Affected individual (define shading in key/legend)	■	●	◆
Affected individual (more than one condition)	▣	◑	◈
Multiple individuals, number known	5	⑤	⟨5⟩
Multiple individuals, number unknown	n	ⓝ	⟨n⟩
Deceased individual	d. 35 y	d. 4 mo	
Stillbirth (SB)	SB 28 wk	SB 30 wk	SB 30 wk
Pregnancy (P)	P LMP: 7/1/94	P 20 wk	P
Spontaneous abortion (SAB)	Male	Female	ECT
Affected SAB	Male	Female	16 wk
Termination of pregnancy (TOP)	Male	Female	
Affected TOP	Male	Female	

Some examples of standardized nomenclature for human pedigree construction. (After Bennett et al., 1995.)

Finn and Smoller (2006) further noted that providing accurate information is important for several reasons. First, in cases in which risk is underestimated, counselors can educate clients about the importance of identifying initial symptoms, thereby facilitating early effective treatment. Second, in cases in which risk is overestimated, counselors can offer reassurance to those individuals who are planning to have children. In addition, relatives who tend to overestimate risk can be cautioned against becoming too vigilant in trying to identify symptoms in low-risk individuals. When family members inadvertently label an unaffected individual as "affected," there is a risk of subjecting that individual

Definitions				Comments
1. Relationship line 3. Sibship line 2. Line of descent 4. Individual's lines				If possible, male partner should be to left of female partner on relationship line. Siblings should be listed from left to right in birth order (oldest to youngest). For pregnancies not carried to term (SABs and TOPs), the individual's line is shortened.
Relationships				A break in a relationship line indicates the relationship no longer exists. Multiple previous partners do not need to be shown if they do not affect genetic assessment.
Consanguinity				If degree of relationship not obvious from pedigree, it should be stated (e.g., third cousins) above relationship line.
Twins	Monozygotic	Dizygotic	Unknown	
No children by choice or reason unknown		Vasectomy	or Tubal	
Infertility		Azoospermia	or Endometriosis	
Adoption	In	Out	By relative	

subsequently presented its peer-reviewed recommendations to the NSGC, which were adopted and published in 1995 (Bennett et al., 1995)

Adherence to a standardized format for recording vital medical information seems a requisite for any health profession. The recent realization of this fact for pedigree charts speaks to the rapidly increasing utility of hereditary information. Whereas such information might once have been of value only to individual practitioners working with specific families, information sharing across practices and research disciplines now requires a more standardized approach to this work.

(usually a child) to unnecessary therapeutic intervention, most notably drug treatment. In addition, such labeling tends to further perpetuate the cycle of family members overestimating genetic effects.

Genetic counselors may also use the results of genetic testing to counsel their clients. In the case of a developing fetus, such testing can be performed by means of an **amniocentesis**, which is the extraction of a small amount of the amniotic fluid that surrounds the fetus in the uterus. This procedure is usually performed between weeks 15 and 20 of gestation, but may be performed as early as week 13. An alternative procedure for collecting cells to be used for genetic testing is

BOX 14.2 Inbreeding (continued)

such a pairing? The probability of a detrimental birth defect of any type (including genetic defects) in the general population ranges from 3% to 5%. For the offspring of two first cousins, that probability is approximately doubled (Zlotogora, 2002).

In North America, there is a frequent misconception that **consanguineous unions** (marriages of blood relatives) are universally forbidden by civil law. In fact, first-cousin marriages are currently legal in 26 of the 50 U.S. states, as well as in Canada, Mexico, and every country in Europe. In other countries (including Qatar, and some regions within Iran), first-cousin marriages are not only legal, but constitute nearly half of all marriages (Saad and Jauniaux, 2002; Saadat et al., 2004). Thus, restrictions on first-cousin marriage are actually the exception to United States law and are very rare on a worldwide basis.

It is likely that some of the effects of inbreeding also reflect assortative mating patterns, particularly in isolated populations. Consider, for example, assortative mating based on low cog-

nitive function. If this mating pattern occurs within an isolated population with a high degree of consanguinity, there is an increased likelihood that the shared pathology stems from shared genetic influences. In such cases, the probability that offspring will show cognitive impairment is increased by assortative mating (through unique combinations of contributing alleles) as well as by inbreeding (through perpetuation of shared contributing alleles). Those who oppose consanguineous unions might support their position by citing the nearly 100% increase in the probability of birth defects seen in first-cousin marriages. On the other hand, those who favor marriage between blood relatives could counter that argument by noting that even this elevated probability of birth defects still indicates healthy birth rates of greater than 90% for such unions (Bennett et al., 2002). What is certain is that marriage among blood relatives is a fairly common practice, and thus a prime example of the need for genetic counseling for educating parents and ensuring optimum health of offspring.

What kinds of recommendations do genetic counselors make?

As noted in the previous section, genetic counselors are often instrumental in guiding pregnant mothers in their decision to undergo an amniocentesis or chorionic villus sampling for genetic screening. Genetic counselors also help parents interpret the results of such tests, explaining how the presence of chromosome or gene aberrations is likely to influence the development of their child. However, these responsibilities typically represent only the starting point for genetic counseling. Once a genetic predisposition has been established, counselors must be prepared to work with clients who will consider the options of continuing or terminating a pregnancy. In cases where parents choose to continue the pregnancy of a child with an apparent genetic defect, the counselor must teach the parents how to take an active role in guiding their child's phenotypic development.

One important application of parent education is to explain how **environmental alterations** can be used to thwart detrimental patterns of development and enhance beneficial patterns. PKU is an excellent example of a disorder

---

TABLE 14.2 Dietary Restrictions for Infants and Children with PKU

FOODS TO AVOID (VERY HIGH IN PHENYLALANINE):

- All foods and beverages containing aspartame (Nutrasweet)

FOODS THAT CAN BE EATEN ONLY IN VERY LIMITED QUANTITIES (HIGH IN PHENYLALANINE):

- Dried beans such as lentils, split peas, garbanzo beans, and kidney beans
- High-protein baked goods
- Eggs
- Meats, fish, seafood, and poultry of all kinds
- Milk, cheese, yogurt, and ice cream
- Nuts, seeds, peanut butter, and other nut butters
- Soybeans, tofu, and other soy products

FOODS THAT CAN BE EATEN IN NORMAL QUANTITIES:

- Certain vegetables, fruits, and fruit juices
- Honey, jam, and jelly
- Ketchup
- Butter
- Low-protein baked goods
- Pancake and fruit syrups
- Pasta, potatoes, and rice
- Protein-free egg substitute
- Soda (not containing phenylalanine)
- Sugar, brown sugar, and molasses
- Wheat starch (used in cooking)

Nursing babies should consume very limited quantities of breast milk or infant formula. Specially designed formulas are available for PKU infants and can be consumed in normal amounts (Lofenalac, PKU1, Phenyl-free).

Source: Adapted from Healthtouch, 2006.

---

that calls for this approach. Recall that PKU is a single-gene recessive disorder. Typically, when a developing fetus is at risk of PKU, at least one parent will be aware of PKU in their family history. In that case, a genetic counselor is likely to recommend some form of fetal cell sampling, which could be used to identify the presence of PKU alleles in the fetus. If it is established that the fetus has two copies of the PKU allele, the parents would be advised by the genetic counselor to implement the strict dietary regimen that can be used to buffer the adverse effects of this disorder (Table 14.2). More specifically, the parents would learn how to restrict their child's intake of phenylalanine through infancy and childhood. To enhance this educational process, the genetic counselor might refer the parents to a dietician for further consultation.

Beyond this specific example of referring parents of children with PKU to a dietician, genetic counselors are a primary resource for a wide range of referrals. Specific organizations, support groups, clinics, and other resources are available to people affected by almost every genetic disorder. Genetic counselors also keep abreast of current research and clinical trials for drug treatment and other therapies that might be of interest to their clients. Finally, these counselors should have a full understanding of in vitro fertilization options, which might be used by couples who are at high risk for conceiving a child with a genetic disorder. For example, when using in vitro fertilization a preimplantation genetic diagnosis can be employed to screen candidate embryos, allowing selection of those that do not have the genetic defect(s) in question. To this end, the field of genetic counseling requires an expansive understanding of legitimate and appropriate resources for a large and growing number of disorders.

Clearly, a person who wishes to pursue a career in genetic counseling must be willing to accept a wide range of responsibilities. Furthermore, the boundaries of these responsibilities are currently expanding at a rate unrivaled by most areas of health and human services. The combination of these factors makes the decision to become a genetic counselor both difficult and exciting. As such, genetic counseling represents a suitable career choice for progressive, dedicated, and highly motivated individuals. More information is provided in Chapter 15 on college degrees and advanced degree programs that cater to this growing field.

Applied Pharmacogenomics

Although genetic counselors might be viewed as a "first line of defense" in guarding against the potential detrimental effects of genetic disorders, unless they possess a medical degree (which some do), they cannot directly prescribe drug therapy. In cases in which genetic disorders are best treated with drugs, a medical practitioner assumes responsibility for drug therapy. Psychiatrists are probably the most common class of physicians treating behavioral pathologies with drugs, but doctors from a variety of specialty areas may prescribe drugs to treat a wide range of genetic disorders. Treatment of degenerative disorders such as Huntington or Tay-Sachs disease, for example, would most likely be assumed by a neurologist. Drugs for children affected by genetic disorders, on the other hand, are frequently prescribed by pediatricians (Figure 14.3). In fact, the vast majority of drug therapy is directed by physicians who do not specialize in treating genetic disorders, but rather have a more basic education in treating the

FIGURE 14.3 The vast majority of drugs used to treat genetic disorders are prescribed by physicians who do not specialize in genetics, such as pediatricians. Drugs are initially prescribed based on assessment of the patient's symptoms. Drug therapy is then guided largely by both the positive and negative effects observed in the patient. Advances in pharmacogenomics will provide new genetic markers that can be used to prescreen candidates for drug treatment, with the hope of reducing treatments that are ineffective or produce adverse side effects. Such benefits will be particularly welcomed by physicians treating children, whose developmental changes make rapid, effective results particularly desirable, and who may be highly distressed by drug side effects.

symptoms that result from gene abnormalities. Advances in pharmacogenomics will benefit a range of medical practitioners by redirecting the current approach of treating symptoms to a more effective approach of treating the primary causes of genetic disorders.

Revisiting basic concepts

With the understanding that a wide range of medical practitioners treat symptoms of genetic disorders, one can begin to appreciate the importance of basic genetic research. The field of pharmacogenomics focuses on the interaction between genes and drug efficacy. Some pharmacogenomics researchers may evaluate the effectiveness of a drug in a large sample of individuals affected by a particular allele. Other researchers in this field may take the more basic approach of determining which proteins are affected by specific genes. Understanding how receptor or transporter proteins are affected by polymorphisms provides valuable information for predicting drug efficacy. Gene mutations that produce receptor proteins with an abnormal structure or function, for example, would presumably cause abnormal responses to drugs designed to bind to and activate those proteins. As discussed in Chapter 3, data from all areas of behavior genetics research have the potential to enhance the development of more effective drugs and drug therapy regimens and to reduce dependence on the trial-and-error approach for treating patients who are in immediate need of effective drug therapy.

Application of current understanding: Schizophrenia

Throughout this text, it has been implied that pharmacogenomic findings can be useful in making decisions about drug therapy for psychiatric disorders. The theoretical basis for pharmacogenomics is certainly sound, yet there remains a question of how effectively its concepts can be applied at our current level of understanding. Although it is progressing at a rapid pace, basic research to identify the specific genes that underlie polygenic disorders is still at a very early stage. Thus, a brief examination of the potential for applying current pharmacogenomic findings to the treatment of complex behavioral disorders might be useful.

Schizophrenia was previously mentioned as a disorder whose treatment could greatly benefit from an understanding of the genetic factors underlying it. There is noticeable variation in patient responses to drug treatments for schizophrenia. These factors, along with its prevalence, severity, and poor prognosis, have put schizophrenia at the forefront of pharmacogenomic research on behavioral disorders.

The first drugs used to treat schizophrenia were developed in the 1950s and are referred to as classical or typical neuroleptics. As described in Chapter 6, these drugs are potent dopamine antagonists that primarily affect the D_2 dopamine receptor subtype. The most profound therapeutic effect in patients who respond positively to this class of drugs is a reduction in positive symptoms. These drugs are less effective at reversing negative symptoms. Typical neuroleptics continue to be a viable treatment option for a large percentage of schizophrenics. However, there are two notable problems with this class of drugs. First, some individuals show little or no decrease in psychiatric symptoms when taking these drugs. Second, even in those individuals who do show improvement, side effects are common, often leading to noncompliance with the treatment regimen.

More recently developed antipsychotic drugs have been designed to exert antagonistic effects at dopamine receptors other than D_2 (i.e., D_1, D_3, D_4) or at serotonin receptors. Drugs from this newer class of antipsychotics are known as second-generation or atypical neuroleptics. They are highly effective in the majority of patients with schizophrenia and tend to reduce both positive and negative symptoms. In addition, atypical neuroleptics produce far fewer and less severe side effects than typical neuroleptics. Although all of these features might appear to make atypical neuroleptics the obvious first choice for treatment of schizophrenia, there is one important caveat. Approximately 1% of patients taking atypical neuroleptics show an adverse response of dramatically decreased white blood cell production, a condition known as **agranulocytosis**. Because agranulocytosis is potentially fatal, patients undertaking atypical neuroleptic treatment must be carefully monitored. This requirement for close monitoring markedly reduces the number of viable candidates for this class of drugs.

Clozapine is currently among the most frequently prescribed atypical neuroleptics, and pharmacogenomic researchers have begun work to determine which patients have the highest probability of responding positively to this particular drug. Before this work began, phenotypic characteristics were the primary variable considered in this decision-making process. Lieberman and colleagues (1994) found that, in general, female patients and patients with early ages of symptom onset were the most likely to respond poorly to clozapine. They also found that clozapine was most effective in patients diagnosed with paranoid forms of schizophrenia. Although such findings are useful, limiting treatment with clozapine, or other atypical neuroleptics, based on these features alone drastically reduces the use of these potentially effective drugs to a relatively small percentage of viable candidates.

More recently, investigators have attempted to find specific alleles that could influence the way in which clozapine and other atypical neuroleptics exert their effects. If such alleles could be identified, they would serve as useful markers for selecting patients most likely to benefit from this class of drugs. In reviewing this topic, Mancama et al. (2002, 2003) found that alleles for dopamine receptors had little predictive value. However, growing evidence suggests that identifying alleles known to influence both serotonin receptors and serotonin transporters may offer valuable insights into which patients might best benefit from atypical neuroleptic treatment. The finding of a link between drug effectiveness and gene abnormalities affecting the serotonin system is interesting for several reasons. First, in terms of basic research, these results strongly suggest a role for the serotonin system in some forms of schizophrenia. This conclusion indicates a need for further research into the influence of serotonin on this pathology, which was formerly associated primarily with dopamine system dysfunction. Second, it is clear that further development and testing of serotonin antagonists is warranted in the search for viable treatment options. Finally, in terms of clinical application, these findings may have revealed the first in a series of genetic markers that can be effectively used to prescreen candidates for neuroleptic treatment. More specifically, the presence or absence of alleles linked to positive responses to atypical neuroleptic treatment could be considered in the decision to prescribe these drugs. Such prescreening would be particularly useful for patients who might otherwise be denied a drug such as clozapine based on their phenotype (e.g., a female with an early age of onset of nonparanoid symptoms). Although this example may seem tentative, it marks an important starting point for the practical use of genetic screening in improving both the safety and the effectiveness of treatment for this most devastating of all psychiatric disorders.

Application of current understanding: Depression

Even though mood disorders generally do not have the profound behavioral consequences seen with schizophrenia, they have become a priority for behavior genetics researchers due to their prevalence. As with schizophrenia, understanding of the pharmacological mechanisms involved in effective treatments for mood disorders has provided insight into their possible genetic underpinnings. Selective serotonin reuptake inhibitors (SSRIs) are highly useful for treating a variety of mood disorders, most notably major depression. As described in Chapter 3 (see Figure 3.16), these drugs have a primary effect of disabling serotonin transporter (SERT) proteins that normally reduce the effects of serotonin.

Although the exact mechanism through which SSRIs exert their antidepressant effects remains a topic of debate, there is little question that genetically influenced alterations in serotonin system function can produce some forms of depression (see Chapter 12). This conclusion would seem to suggest that the identification of particular alleles affecting the serotonin system could be used to predict specific system alterations responsible for psychopathologies. Such an in-depth understanding of the neurochemical underpinnings of mood disorders would be a marked advance in the process of selecting initial drug treatment. At this time, the integration of mood disorder research in genetics, neurochemistry, and pharmacology is far from complete. However, some information from this area is noteworthy.

In a recent review of pharmacogenomic research on major depression, Serretti and colleagues (2005) were cautiously optimistic about the prospects of using genetic screening to guide drug treatment of mood disorders. As an example, they cited mixed findings from genetic screening research on patient responses to SSRIs. Several studies in this area have determined that patients with depression who possess a "long" allele of the serotonin transporter (*SERT*) gene respond much better to fluvoxamine (an SSRI) than do patients who possess a "short" allele of the same gene. One problem with these results, which were generated from a sample composed largely of Caucasian patients, is that they contrast with at least two studies that found the opposite to be true. More specifically, when Asian patients were studied, the results showed a more positive response to fluvoxamine when the "short" allele was present. Serretti and colleagues recognize that such conflicting results tend to raise skepticism about the validity and reliability of this approach. However, they also note that even though methodological problems can create such discrepancies, there is also a possibility that ethnicity plays a role in gene expression. If that is indeed the case, such results show an even greater potential role for genetic screening than originally envisioned.

Clearly, much more pharmacogenomic research on mood disorders is needed. These early studies, however conflicting or inconclusive, are laying important groundwork upon which future researchers can build. Recall that Mendel's early findings could not adequately explain much of the variation seen in hereditary patterns, but his essential basic research produced the foundation for our current, highly complex understanding of gene transmission. It is hoped that pharmacogenomics has the potential to develop into a viable and practical area of research with direct clinical applications in the near future.

Gene Therapy

Information on the use of gene therapy has long been a staple of genetics and heredity textbooks. This information, however, has been relegated largely to discussions of "future perspectives" on applications of genetic research. Although this text also reserves some discussion of gene therapy for the final chapter on future perspectives, a growing body of literature warrants its inclusion in the present chapter. In particular, some basic research findings are now being assessed for their therapeutic potential in human trials.

Genetic engineering is a term used to describe the process of inserting new genetic information into a cell with the intention of changing that cell's functions and ultimately modifying the phenotype of the organism. The term *genetic engineering* is sometimes used synonymously with the term *gene therapy*. In fact, **gene therapy** is a specific application of genetic engineering techniques with the primary objective of correcting defective genes to treat a genetic disorder. Thus, techniques developed in the general field of genetic engineering can be used for purposes ranging from creating disease-resistant strains of fruit to developing treatments for neurodegenerative diseases. The focus of this section is research on potential applications of gene therapy to curing or preventing human disease, with a particular emphasis on psychiatric diseases and other types of neurological dysfunction.

In theory, the basic concepts that underlie gene therapy are as follows:

1. Cells containing dysfunctional DNA are removed from the organism.

2. The dysfunctional DNA is altered by inserting functional DNA sequences from another organism or by inserting DNA sequences that have been manipulated using laboratory techniques.

3. The newly altered DNA must be viable, functional, and capable of replication.

4. Cells containing the newly altered DNA are reintroduced into the organism, typically into the organ where they are intended to function.

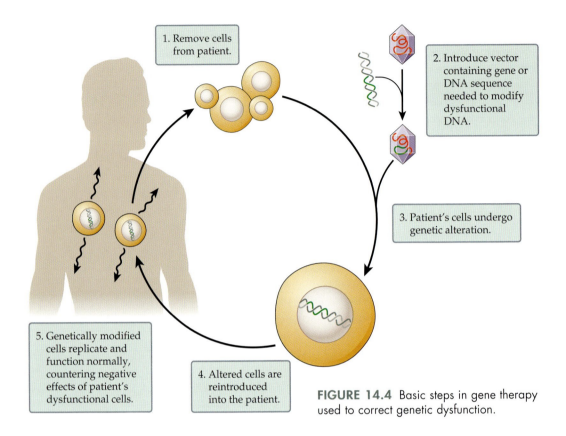

1. Remove cells from patient.

2. Introduce vector containing gene or DNA sequence needed to modify dysfunctional DNA.

3. Patient's cells undergo genetic alteration.

4. Altered cells are reintroduced into the patient.

5. Genetically modified cells replicate and function normally, countering negative effects of patient's dysfunctional cells.

FIGURE 14.4 Basic steps in gene therapy used to correct genetic dysfunction.

5. The cells containing the newly altered DNA replicate and function normally, counteracting or (ideally) completely overriding the negative effects of the organism's dysfunctional cells.

The evolution of these concepts toward practical applications (Figure 14.4) has unfolded rapidly and dramatically over the past century, creating a field with a brief yet eventful history and an even more intriguing future.

Historical perspective

Friedmann (1992) provides an excellent review of early research in the area of gene therapy, with a focus on particular milestones in the field. In that review, it is revealed that conceptual theories of genetic engineering were published as early as the 1940s. Testing of those theories followed in the 1960s, when several research groups reported that exogenous DNA could be permanently integrated into, and subsequently expressed by, mammalian cells. Direct application of gene therapy concepts is apparent in at least two of these early genetic engineering studies, which targeted cancer cells for manipulation (Rabotti, 1963; Majumdar and Bose, 1968).

One noteworthy obstacle encountered in early genetic engineering research was the lack of an effective method for delivering new DNA into target cells. By the late 1960s, several laboratories had reported the successful use of viruses as delivery vehicles, or **vectors**, for DNA. A virus is a parasitic infectious agent, capable of replicating only when integrated into a host cell. Viruses were recognized as an obvious choice by early researchers because of their ability to enter a host cell without producing immediate damage to that cell.

Throughout the 1970s, continued assessment of desirable characteristics for vectors led researchers to investigate the use of **retroviruses**, a group of viruses that contain RNA (rather than DNA) as well as an enzyme capable of translating that RNA into DNA. Translating RNA into DNA should sound counterintuitive, as you learned in your introductory biology course that DNA codes for RNA and not vice versa. In fact, the prefix *retro* is a reference to the unique ability of these viruses to create DNA from RNA, a process referred to as **reverse transcription**. The best-known example of a retrovirus is probably the human immunodeficiency virus (HIV), which most frequently infects cells of the immune system. HIV, like other retroviruses, uses its unique enzyme, called reverse transcriptase, to make a DNA copy of itself, which it then integrates into the host cell's DNA (Figure 14.5). The host cell is thus forced to use its own DNA transcription process to produce new viral RNA. The resulting new copies of HIV are eventually released from the host cell to infect and repeat the process in additional host cells.

Although HIV, which destroys the immune system cells it infects, probably does not conjure images of a therapeutic tool, consider its general features for a moment. A retrovirus is capable not only of entering a host cell with minimal damage, but also of altering the DNA of its host. This alteration, in turn, leads to a stable change in gene-controlled cell function. Gaining an understanding of the mechanisms of retroviruses, combined with advances in techniques that allowed the insertion of particular base sequences into retroviruses, led to the creation of retroviral vectors that are, in theory, ideal for gene therapy. Use of retroviruses for altering human DNA began in earnest in the 1980s, and they remain the most viable vectors for this purpose today.

Applying current knowledge to behavior genetics

The overwhelming majority of genetic engineering work has been conducted using nonneural tissue. Research on the use of gene therapy to redirect neural structure and function is in the very early stages. Most such research is being done in animal models, with only a small number of human trials contributing

FIGURE 14.5 The life cycle of HIV illustrates the basic mechanisms of retroviral replication.

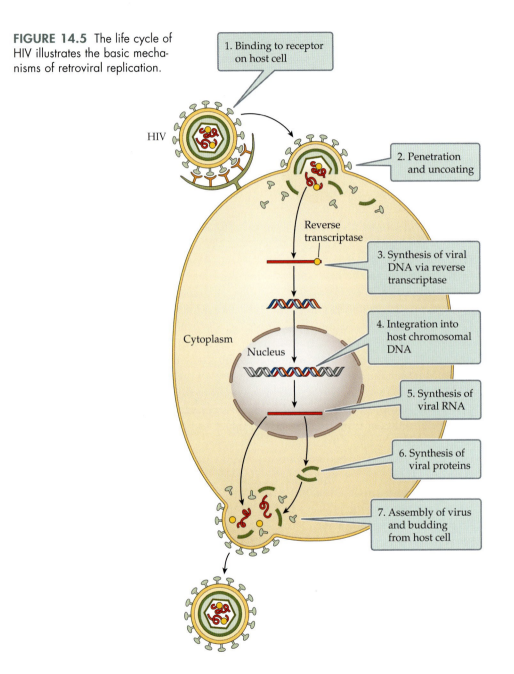

1. Binding to receptor on host cell

HIV

2. Penetration and uncoating

Reverse transcriptase

3. Synthesis of viral DNA via reverse transcriptase

Cytoplasm

Nucleus

4. Integration into host chromosomal DNA

5. Synthesis of viral RNA

6. Synthesis of viral proteins

7. Assembly of virus and budding from host cell

to the literature. In a recent review, Sapolsky (2003) nicely summarized the current advances in this area, placing an emphasis on the potential practical applications of findings from preliminary studies. One notable feature of Sapolsky's review is his subdivision of this research into two distinct categories. The **neurological realm** of gene therapy encompasses research into methods of slowing the progression of cell loss caused by acute neurological insult (e.g., stroke) or chronic neurodegenerative diseases (e.g., Alzheimer disease, Parkinson disease). In contrast, the **psychiatric realm** of gene therapy can be defined as research designed to develop treatments for more subtle, less clearly defined neural dysfunctions associated with psychiatric disorders (e.g., schizophrenia, autism, affective disorders).

Up to this point, research on the use of gene therapy to alter neural function has focused primarily on the neurological realm. In part, this emphasis has resulted from our somewhat better understanding of neurological dysfunction than of psychiatric disorders. In addition, the urgency of discovering effective treatments in the neurological realm is perceived to be greater. For example, neurological gene therapy has the potential to prevent the massive, potentially lethal, damage produced by acute neurological insult as well as the debilitating effects of chronic neurological diseases, which often lead to loss of independence and premature death. In contrast to these lethal outcomes, the symptoms of many psychiatric disorders can often be reduced to tolerable levels with drug therapy.

The starting point for any gene therapy is to identify a DNA sequence with the ability to reverse endogenous dysfunction. In cases in which a genetic mutation causes abnormal neural structure or function, one solution would be to introduce nonmutated DNA sequences (representing normal genes) into the system. Once integrated into target cells, the exogenously developed gene would be integrated into the system through the normal process of replication, ideally normalizing the function of affected tissues or organs. A second viable approach to gene therapy is the introduction of an allele designed to be more or less active than the normal allele. For example, in cases in which chronic cell loss underlies symptoms (as in neurodegenerative disorders), introduction of alleles designed to overproduce neuronal growth factors in specific brain regions might have the potential to slow cell loss. More details on specific forms of gene therapy are discussed in Chapter 15, the final chapter of this text.

Chapter Summary

Several notable events have shaped the field of genetic counseling. The founding of the National Society for Genetic Counselors (NSGC) in 1979 was followed two years later by the development of the American Board of Medical Genetics (ABMG), a sanctioning body for certifying genetic counselors. The *Journal of Genetic Counseling* was first published in 1992, and, in 1993, the rapidly expanding ABMG was restructured into the American Board of Genetic Counselors (ABGC), which now certifies genetic counselors and accredits genetic counseling programs. Most recently, in an effort to provide a concise and accurate description of the field, the NSGC adopted a formal definition of genetic counseling.

When seeking genetic counseling, clients provide information about their family history of physical or psychological dysfunction. This information is used to counsel and educate clients, who frequently over- or underestimate inheritance risks. Clients can then make better-informed decisions about having children, or about treating children who are symptomatic for a particular disorder.

Some genetic counselors visually inspect karyotypes to identify a wide range of chromosomal abnormalities, including trisomies, monosomies, or structural defects, such as breaks, fragments, and translocations. Such analysis requires familiarity with the normal degree of bending, overlapping, stretching, and chromosome rotation that is expected in karyotypes but does not indicate true genetic anomalies. Counselors can also recommend screening for known alleles associated with single-gene disorders such as PKU, cystic fibrosis, Tay-Sachs disease, sickle-cell anemia, and Huntington disease. Karyotype analysis and genetic screening can be conducted for a developing fetus by means of an amniocentesis or chorionic villus sampling. In the case of polygenic disorders, genetic counselors must evaluate pedigrees not only for the pathology of interest, but also for related traits associated with elevated risk. In some cases, the influence of assortative mating must be considered.

Patient education is an essential role of genetic counselors. Such education includes explaining to clients how environmental alterations can be used to reduce the negative effects of genetic influences. Genetic counselors also direct individuals to organizations, support groups, clinics, and other resources that can help them. In some cases, genetic counselors may refer clients for clinical trials of experimental drug treatments or other newly developed therapies.

The field of pharmacogenomics is concerned with understanding the interaction between genes and drug efficacy. Schizophrenia was originally treated with typical neuroleptics (antagonists that act primarily at D_2 dopamine receptors), which effectively block positive symptoms in only a subpopulation of patients and also tend to produce undesirable side effects. A newer class of drugs, called atypical neuroleptics, blocks serotonin receptors as well as dopamine receptors other than the D_2 subtype. Atypical neuroleptics reduce both positive and negative symptoms in most patients and produce fewer side effects than typical neuroleptics. However, in a small percentage of individuals, their use can result in agranulocytosis, a potentially fatal side effect. Researchers have recently found variations in genes controlling serotonin systems that may help predict patient responses to atypical neuroleptics. This finding is important for several reasons. First, it implies a primary role for serotonin instead of, or in addition to, dopamine in some forms of schizophrenia. Second, it suggests a need for further research exploring serotonin drug treatment for schizophrenia. Finally, it indicates that methods may soon be available for prescreening candidates for more specific selection of neuroleptic treatment.

Currently, SSRIs are among the drugs most commonly prescribed to treat mood disorders, particularly major depression. There is growing optimism that screening for gene anomalies linked to serotonin systems may soon be used in the process of selecting antidepressant drug treatments. More specifically, one particular *SERT* allele has been associated with a positive response to fluvoxamine (an SSRI), whereas a second *SERT* allele has not. Interestingly, this finding, which came from a sample of Caucasian individuals, could not be replicated in Asian patients. One possible reason for this discrepancy is the role of ethnicity in gene expression.

Genetic engineering is the process of altering genetic information within a cell with the intention of changing cell function and ultimately modifying the phenotype of an organism. Gene therapy is a specific application of genetic engineering designed to correct defective genes and treat genetic disorders. In simple terms, this is accomplished by removing cells from an organism and inserting viable, functional, and corrective DNA sequences into those cells. Cells with corrected DNA are then reintroduced into the organism with the hope that they will replicate and counteract the negative effects of dysfunctional endogenous cells. Although conceptual theories of genetic engineering were published as early as the 1940s, one obstacle in its practical application was the lack of an effective method of delivering DNA into target cells. This problem was resolved with the discovery that retroviruses are highly effective vectors for transporting modified DNA.

Most genetic engineering work has been conducted using nonneural tissue. Few studies have focused on the use of gene therapy to redirect neural structure and function, and most such research has been in animal models. Gene therapy designed to alter neural structure and function can be categorized as either neurological or psychiatric. Neurological gene therapy encompasses methods of slowing the progression of cell loss caused by acute neurological insult or by chronic neurodegenerative disorders. Psychiatric gene therapy encompasses treatment of the more subtle, less clearly defined neural dysfunction associated with psychiatric disorders. Up to this point, research has focused primarily on the neurological realm.

15

The Future of
Behavior Genetics

This text has presented a wide range of information on behavior genetics, with an emphasis on historical developments and the current state of the field. Throughout the previous chapters, some implications for future applications of these research findings have been integrated into the discussion. However, the future of behavior genetics will be vast, and thus a more detailed look at some particular aspects is needed. Keep in mind that the information in this chapter is fluid, such that with each passing year, speculations will become the realities of behavior genetics, and those realities will, in turn, fuel new speculations. This, of course, is the optimistic cycle for any fledgling scientific field. More specifically, one can foresee a larger role for genetic counseling as scientists unravel the masses of information collected by the Human Genome Project and other related research. From these data, a clearer understanding of environmental influences on genetic predisposition will undoubtedly emerge. Genetic counselors will be the front line practitioners disseminating this information. As methods for altering DNA are elucidated and refined, it is likely that gene therapy will be gradually integrated into therapeutic approaches to treating illnesses currently defined as "untreatable" or "incurable." Finally, as a necessary by-product of this growing knowledge, educational programs will be required to foster understanding among the general public, as well as to train those wishing to specialize in fields related to behavior genetics.

Reviewing the Past with an Eye to the Future

The strides made in the field of behavior genetics, particularly over the last 20 years, are likely to affect many related areas of research in the near future. There is a sense of excitement that comes with living during a time in history that parallels the genesis of such important experimental findings. To witness this explanation of genetics unfolding, and to see the potential it has to enhance our understanding of the physiological basis for such a wide array of behaviors, is truly a remarkable experience. At the same time, our optimism and enthusiasm must be tempered by the realization that the practical application of these findings is a highly complex undertaking that will require tremendous effort and, of course, much more time.

Future applications of the Human Genome Project

The primary goals outlined for the Human Genome Project were completed in 2003. Most notably, that year marked the time when 99% of the gene-containing regions of human DNA had been sequenced with a level of 99.99% accuracy. In that same year, the total number of known **single nucleotide polymorphisms**, or **SNPs** (pronounced "snips"), reached 3.7 million. As the name implies, a SNP is a single nucleotide (A, T, C or G) variation between DNA sequences taken from different members of the same species (Figure 15.1). The vast majority of

Unaffected:	AATCGACCGATTCAGCTTCCCCAGTT
Unaffected:	AATCGACCGATTCAGCTTCCCCAGTT
Affected:	AATCGACCGATTCAGCTTCACCAGTT
Unaffected:	AATCGACCGATTCAGCTTCCCCAGTT
Affected:	AATCGACCGATTCAGCTTCACCAGTT

FIGURE 15.1 SNPs (as the name implies) are alterations in a single nucleotide. In cases where SNPs occur within a DNA coding sequence, these minute differences may contribute to phenotypic variation. Thus, by comparing groups of affected and unaffected individuals, researchers may be able to identify SNPs that are correlated with susceptibility to genetically influenced diseases and then use these SNPs to identify individuals who may be susceptible to these disorders.

SNPs occur in noncoding DNA sequences. These variations have no functional consequences, but may hold value as markers for larger polymorphisms or may be used in DNA fingerprinting. However, SNPs that occur within a DNA coding sequence (i.e., within a gene) may contribute to phenotypic variation. Thus, researchers frequently compare the SNPs of groups of affected versus unaffected individuals in their search for genetic contributions to a trait of interest (most notably disorders with an inheritance pattern suggesting the involvement of genetic factors). If these comparisons find SNPs that are correlated with genetically influenced diseases, researchers may then be able to use those SNPs to identify individuals who may be susceptible to these disorders.

In time, and with the accumulation of a large database indicating where nucleotides tend to vary between affected and unaffected individuals, SNPs could become a useful tool for genetic screening. Such screening would have obvious benefits in the area of pharmacogenomics, as it would allow physicians to begin drug treatment for a genetic disorder prior to expression of the phenotype. In particular, the use of neuroprotectants to prevent cell loss caused by neurodegenerative diseases would be far more effective if applied in a preventive manner, since these drugs do not aid in repair or replacement of cells that have already been affected by the disease. SNP screening may also become a useful tool for genetic counselors. Advice based on the results of SNP screening could include recommendations for pharmacological intervention or dietary restrictions, as well as cautions about susceptibility to environmental factors such as stress, alcohol, or drugs. Ideally, this application of information from the Human Genome Project will eventually allow many more individuals to live healthier lives, armed with the knowledge of their predispositions to particular maladies.

The future of genetic counseling

As indicated above, the field of genetic counseling will need to be highly responsive to future research findings. The basic premise of genetic counseling is that most individuals who request information about genetic influences on their lives (or the lives of their family members) can benefit from the help of a professional who educates and guides them through the decision-making processes associated with genetic testing and treatment of genetic disorders. Such decisions may include whether or not to undergo genetic testing as well as how best to respond when test results indicate a predisposition for a genetically influenced disorder (Figure 15.2). This basic description of genetic counseling has changed little over time. What has changed quite dramatically, and will continue to change, is the volume and complexity of the information being interpreted. The future will undoubtedly hold subfields of genetic counseling, in-

FIGURE 15.2 The Human Genome Project and related research have provided a wealth of new information useful for genetic counseling. However, this increase in the volume and complexity of relevant data demands greater efforts on the part of counselors to accurately interpret and disseminate that information. It is likely that in the future, genetic counselors will need to specialize in particular subfields within the discipline in order to keep pace with advances in basic research.

cluding, for example, counselors specializing in single-gene disorders, polygenic degenerative disorders, polygenic mood and personality disorders, and so on. Within each area of specialty, counselors will continue to disseminate information about inheritance probabilities and environmental factors that contribute to trait expression. However, more preventive treatment options will be available for consideration, including presymptomatic pharmacological interventions (e.g., neuroprotectants) and gene therapy. Just as genetic counselors are now expected to advise clients on the risks and benefits associated with conceiving a child or undergoing an amniocentesis or chorionic villus sampling, counselors in the future will be expected to guide an increasing number of individuals in making decisions about therapeutic interventions.

The information available from genetic testing is increasing rapidly, but so are ethical questions about what to do with that information. Csaba and Papp (2003) recently reviewed the ethical issues that currently face genetic counselors and speculated on some future concerns. They noted that results from the Human Genome Project have led to a rapid increase in both the range and the accuracy of two testing paradigms in particular (Table 15.1). The first, **presymptomatic testing**, is used when healthy individuals are assessed for an allele that is destined to produce a disease state, such as Huntington disease or early-onset Alzheimer disease. The second, **susceptibility testing**, is used when individuals are assessed for inheritance of multiple alleles that put them at risk for a disease, such as late-onset Alzheimer disease or schizophrenia. The identification of an increasing number of single-gene disorders, as well as information detailing the contributions of individual genes to polygenic disorders, has made both presymptomatic and susceptibility testing more relevant and useful for genetic counselors. At the same time, interpretation of the information

TABLE 15.1   Basic Features of Presymptomatic Testing and Susceptibility Testing

	PRESYMPTOMATIC TESTING	SUSCEPTIBILITY TESTING
Conducted in:	Healthy individuals	Healthy individuals
Used to identify:	Alleles that are certain to cause dysfunction if the individual lives long enough	Alleles that are associated with an increased risk of dysfunction
Genetic basis:	Single gene	One or more genes
Examples:	Huntington disease	Cancer, heart disease
	Early-onset Alzheimer disease	Late-onset Alzheimer disease
Treatment options:	Very limited at this time	Pharmacological interventions
		Environmental interventions (e.g., diet, exercise, stress reduction)

Source: Csaba and Papp, 2003.

gathered from such tests is growing increasingly complex as the volume of relevant research continues to grow.

Pregnancy termination by induced abortion is a topic that arouses a great deal of emotionally charged controversy (Figure 15.3). One concern about recent advances in the identification of genetic aberrations in utero is that such information could lead to an increase in requests for abortions. Currently, amniocentesis and chorionic villus sampling can be used to detect gross chromosomal abnormalities and the presence of single-gene disorders in developing fetuses. In some cases, that information is used as a basis for choosing to terminate a pregnancy. This is particularly true in the case of trisomies, for which rates of voluntary pregnancy termination are typically above 80% (for review, see Roberts et al., 2002). Thus, opponents of voluntary abortion might also express opposition to advances in genetic research that could be used in genetic counseling. However, there are at least two basic reasons why such opposition may be misdirected.

First, a primary goal of research in the field of behavior genetics is to establish viable treatments or preventive measures for genetic disorders. Achievement of these goals would, in theory, reduce the demand for abortions. Evidence for this potential effect comes from the results of a survey in which individuals were asked about the likelihood of terminating a pregnancy if genetic testing had established that their child would develop bipolar disorder (Smith et al., 1996). When the disorder was described as "severe," involving a high risk of suicide or hospitalization, 75% of respondents indicated that they would opt for

TABLE 15.2 Some Current Problems with Application of Stem Cell Therapy and Some Potential Genetic Engineering Solutions

PROBLEM	POTENTIAL SOLUTION
Implanted stem cells may not reproduce in sufficient numbers to alter function of endogenous dysfunctional cells	Modify genetic makeup of stem cells to increase their replication efficiency or increase their influence on the system compared with endogenous dysfunctional cells
Stem cells from foreign sources may be disabled or destroyed by normal immune system function in the recipient	Modify genetic makeup of stem cells to increase resistance to immune system or to more closely match stem cells to recipient, making them less likely to evoke an immune response
Stem cells may mutate into nonfunctional cells that replicate rapidly, producing cancers	Include a conditional "suicide gene" that becomes active when cell replication exceeds acceptable levels

Source: Strulovici et al., 2007.

risk of introducing novel heritable side effects, a risk not found in somatic cell gene therapy. When we consider the irreversible nature of this procedure, it is difficult to justify such a risk of harm that could be passed on to future generations. Nielsen (1997) nicely summarized many of the technological and ethical concerns associated with pursuing human germ cell gene therapy. Among the primary considerations is the inevitable need for initial human trials to test this highly risky procedure. As Neilsen points out, the medical and ethical risks of somatic cell gene therapy are similar to those currently accepted in organ transplantation. Germ cell gene therapy, on the other hand, requires consideration of a unique ethical question: the potential results in future generations affected by the procedure.

Although germ cell gene therapy may not be a viable technique for quite some time, we must give thoughtful consideration now to the implications of such technology beyond the elimination of genetic disorders. Think, for example, about the possibility of using this technology to increase intelligence, physical strength, emotional stability, and other polygenic traits. Recall the discussion in Chapter 1 of this text of the dark era of eugenics in the history of behavior genetics. Just as it was once proposed that selective breeding should be used to create healthier future generations of humans, germ cell gene alteration could also be used in this way. Although the ability to manipulate complex traits through genetic engineering may never materialize, the possibility does raise questions that should be addressed long before the technology reaches anywhere near that level. After all, it is unlikely that early medical researchers could have imagined routine use of radiation therapy, laser surgical instruments, internal imaging devices such as MRI, or any number of other techniques now considered routine. History has taught us that advances in medical technology can occur at a rapid pace.

One additional innovation in gene therapy research worth mentioning is the potential use of **small interfering RNA (siRNA)** to suppress gene expression. In basic terms, siRNAs are short artificial segments of double-stranded RNA designed to bind to specific mRNA molecules. These siRNAs may be introduced into an organism using a vector in a manner similar to that described for gene therapy. Once incorporated into the cell, the double strand of siRNA binds to a cellular protein called the **RNA-induced silencing complex (RISC)** (Figure 15.5). The protein unwinds the siRNA and guides it to its target mRNA sequence, to which it binds. The presence of a double-stranded sequence containing mRNA stimulates the release of a "slicer" enzyme from the RISC. This enzyme cleaves the mRNA into nonfunctional pieces, halting its translation. This technique has been used for several years in basic research to "silence" specific mRNAs as a means of determining the functional consequences of gene expression. However, researchers have now begun to consider the viability of using siRNA in applied research. Koutsilieri and colleagues (2007) recently reviewed possibilities for using siRNA as a method for suppressing the expression of dysfunctional alleles. Although the concept of using siRNA in a therapeutic role is sound, much additional research is needed before this technique can be applied in practice.

The Future of Behavior Genetics Education

Behavior genetics is a rapidly expanding field that has the potential to affect many related areas of health and human services. Although awareness of behavior genetics research has certainly been raised by the success of the Human Genome Project, most people still possess only a limited understanding of how this type of research will affect them. Therefore, it is worth describing how the general public is being educated to raise its awareness of behavior genetics. It also makes sense to review some of the ways in which interested individuals may pursue a formal education, or a professional career, in behavior genetics.

Educating the public

In the process of advancing the field of behavior genetics, a primary objective will be to educate the public, particularly with regard to how research findings are being applied by health professionals. It will be particularly important to raise public awareness about the usefulness and availability of genetic coun-

FIGURE 15.5 By designing siRNAs with a specific base sequence, researchers can inhibit translation of specific mRNAs. This technique is currently used as a basic research tool, but could have value as a therapeutic approach to blocking the effects of disease-causing alleles.

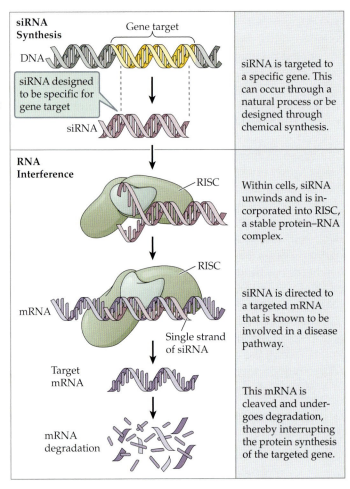

seling. In addition to simply increasing the number of individuals who seek such counseling, there is a more specific need to encourage women to seek counseling prior to pregnancy, rather than waiting until after they become pregnant. Aalfs and colleagues (2007) recently reported that among women seeking counseling in their department of clinical genetics, between 10% and 20% schedule their first visit after becoming pregnant. These researchers pointed out that some disadvantages to waiting until after conception include limiting options for preconception genetic testing and preventive measures, as well as eliminating the possibility of using alternative methods of reproduction (such as artificial insemination or in vitro fertilization). As genetic counselors expand their range of services, other medical practitioners serving women of childbearing age (most notably those practicing obstetrics and gynecology) will be more likely to make referrals to these counselors for patients who indicate a desire to become pregnant.

Although expansion of genetic counseling has the potential to enhance health practices worldwide, it could have particular benefits in cultures with high rates of consanguineous marriages. For example, a recent study by Al-Gazali (2005) found that in the United Arab Emirates, most parents of children with genetic disorders attributed the malady to "God's will," even after having undergone genetic counseling. However, when pressed for alternative explanations, over half of the parents acknowledged a possible genetic basis for the disorder, even though they tended to exhibit an incomplete understanding of the genetic explanation. Al-Gazali suggests that these parental responses demonstrate the need for genetic counseling that includes clearer explanations of genetic information particularly that associated with genetic mechanisms and mathematical probabilities. In addition, this finding suggests that genetic counselors will need to adjust their clinical approaches to the particular cultural and religious beliefs of their clients. In fact, as discussed in the next section, some programs offering advanced degrees in genetic counseling have integrated courses into their curriculum that address such issues. Clearly, the heterogeneity of the world's cultures will require that a vast range of approaches be used in educating the public about these complex biological issues that are so often closely interwoven with societal norms and religious beliefs.

Advanced degrees in genetic counseling

The American Board of Genetic Counseling (ABGC) currently certifies genetic counselors and grants accreditation to graduate programs offering degrees in genetic counseling. Over the past decade, the number of genetic counselors recognized by the ABGC has increased approximately fourfold to a current number of over 2,000. This expansion demonstrates the growing desire for specialists in this field.

Certification by the ABGC first requires that the applicant graduate from an ABGC-accredited master's-level program. At the time of this publication, there were 29 such programs distributed throughout 19 U.S. states and the District of Columbia, as well as 3 programs in Canadian universities (Figure 15.6). Considering the growth of this field, that number is likely to increase steadily over the next several decades. Once a graduate degree in genetic counseling is earned, applicants must then meet additional rigorous criteria, including an ABGC review of graduate school transcripts, evidence of training in a clinical setting, and letters of recommendation. Finally, applicants must complete a comprehensive certification examination developed and administered by the ABGC.

TABLE 15.3 **Practice-Based Competencies Emphasized in ABGC-Accredited Graduate Programs** *(continued)*

DOMAIN IV: PROFESSIONAL ETHICS AND VALUES

1. Can act in accordance with the ethical, legal, and philosophical principles and values of the profession.
2. Can serve as an advocate for clients.
3. Can introduce research options and issues to clients and families.
4. Can recognize his or her own limitations in knowledge and capabilities regarding medical, psychosocial, and ethnocultural issues and seek consultation or refer clients when needed.
5. Can demonstrate initiative for continued professional growth.

Source: American Board of Genetic Counseling, 2006.

counseling programs. Several courses appear consistently across most programs, including basic courses in human genetics, medical genetics, and clinical genetics. All programs also include several courses designed to teach specific skills that a counselor requires, such as General Counseling Communication, Genetic Counseling, Working with Support Groups, and Basic Interviewing Skills. All programs also require some form of clinical practicum, in which the student gains experience in a real counseling situation, usually within a genetic counseling clinic affiliated or collaborating with the university.

Although all of the accredited programs are similar in terms of teaching the basic competencies required for genetic counseling, there are important differences among programs that allow students to tailor their education to a specific career path or specialty area. For example, some programs require active involvement in genetic laboratory research, emphasizing basic methods for understanding the foundations of genetics. Some programs also require courses focusing on topics such as genetic influences on cancer, ethical considerations in genetic counseling, and cultural and religious considerations in genetic counseling. Students who meet the requirements for acceptance into more than one graduate program would be wise to examine the course offerings for each program. By selecting the curriculum that most closely matches their individual interests, students will have taken a vital first step in establishing an interesting and gratifying career in the field of genetic counseling.

The ABGC Web site (www.ABGC.net) is an excellent resource for identifying accredited genetic counseling programs and for learning more about certification requirements. This site also contains a wealth of related information, including a schedule of relevant conferences, links to other resources, and listings of certified genetic counselors in particular geographic regions.

Advanced degrees in behavior genetics

Over the past several decades, genetic counseling has evolved into a relatively specific discipline, with clinicians now educated in accredited programs and certified according to a common standard set by the ABGC. In contrast, the field of behavior genetics remains largely a collage of scientists from a variety of research areas, most notably subfields within biology, psychology, and chemistry. This diversity does not imply a lack of organized efforts among these researchers, however. The field of behavior genetics, as should be apparent after reading this text, covers a very wide range of research specialties, from manipulating the molecular structure of DNA to studying heritable traits in twins and adopted children. It is unreasonable to expect anything less than an expansive array of contributing disciplines for such diverse subject matter. However, there is a common thread connecting these realms of research. That thread, which draws scientists together in a concerted and productive effort, is a shared interest in understanding the specific effects of genes on behavioral differences. The result of these researchers' efforts has been the genesis of the field of behavior genetics. Furthermore, behavior genetics is a field in which technological advances seem to allow these researchers to constantly add new information and, in some cases, challenge them to reevaluate long held beliefs (Box 15.2).

Students wishing to contribute further to behavior genetics may chose from a wide range of fields including, but not limited to, genetics, experimental psychology, biopsychology, neuroscience, biology, molecular biology, statistics, organic chemistry, and general chemistry. In each of these fields, and others, research essential to understanding behavior genetics is constantly being generated.

But what of students who wish to focus specifically on behavior genetics as a field of research? What if a student wishes to evaluate the larger overarching questions, rather than focusing on a specific realm that contributes to this big picture? For these individuals, programs have begun to emerge that offer a specific graduate degree in behavior genetics. Perhaps the most notable example of such an institution is the Institute for Behavioral Genetics (IBG) at the University of Colorado at Boulder. Students from a range of related disciplines can participate in IBG graduate training within the University's interdisciplinary certificate program, which allows them to earn a Ph.D. in their area of choice while enjoying the specialized training offered by the IBG. In addition, the IBG offers two curriculum options (a psychology track and a neuroscience track) for a Ph.D. in the area of behavior genetics. The University of Minnesota also now grants a Ph.D. in behavioral genetics to students who select that track from the broader graduate program in Personality, Individual Differences, and Behavior Genetics. Common elements in the curricula of behavior genetics programs include courses in quantitative and molecular genetics as well as a required laboratory research element.

BOX 15.2 The "Newest" Type of Twin

When people think of the future of genetics, they typically conjure images of genetic engineering, cloning, DNA profiling, and the like. Less frequently do they consider that advances in genetic research can stimulate not only new discoveries, but also the reevaluation of long-held concepts. Several examples were discussed in Chapter 5, where research advances were noted to have revealed numerous exceptions to Mendel's original laws of heredity. But recently, a simpler, even more basic premise of genetics has been the subject of scrutiny: the concept that there are only two types of twins.

As described in Chapter 8, dizygotic twins result from the fertilization of two separate eggs by two separate sperm cells. In contrast, monozygotic twins result when a single egg, fertilized by a single sperm, divides into two distinct embryos early in the process of cell replication. This division, of course, produces two fetuses with identical genomes. However, in 2007, a group of researchers discovered a case in which a third developmental process had produced a unique type of twins (Souter et al., 2007). In this case, it is believed that a single egg began a process of **parthenogenic activation**, starting to divide prior to fertilization. In the midst of this process, the egg was fertilized by two separate sperm (one with an X chromosome, one with a Y chromosome). At that point, cells from the two separately fertilized (but still connected) eggs were intermingled before they eventually separated and began to develop as separate embryos. The end result of this process was the development of twins who shared a noticeably higher percentage of alleles inherited from their mother (from the same egg) than from their father (from the two sperm). In short, the genetic similarity between the twins in terms of alleles inherited from their mother resembled what is expected in monozygotic twins, whereas their genetic similarity in terms of alleles inherited from their father resembled what is expected in dizygotic twins (or nontwin siblings). The researchers concluded that these twins were more genetically similar than dizygotic twins, but less similar than monozygotic twins. Although this phenomenon is probably extremely rare, the fact that advances in genetic research have led to the rethinking of a process as basic as twinning speaks to the potential for future discoveries in the field.

The "three-gamete mechanism" proposed to explain the development of a unique, previously unknown type of twins. (Adapted from Souter et al., 2007.)

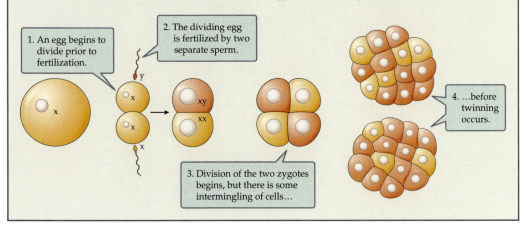

1. An egg begins to divide prior to fertilization.

2. The dividing egg is fertilized by two separate sperm.

3. Division of the two zygotes begins, but there is some intermingling of cells...

4. ...before twinning occurs.

Behavior genetics research facilities can also be found at a number of universities that do not offer a specific degree in behavior genetics. However, these facilities bring together a wide range of faculty members with expertise in diverse areas who can train graduate students wishing to focus their efforts on behavior genetics. The University of Pennsylvania, for example, has a well-established behavior genetics research laboratory used to train students seeking a Ph.D. in gene therapy, genetics and gene regulation, or genomics and computational biology.

An ever-increasing number of collaborations among scientists who contribute to behavior genetics will undoubtedly lead to more behavior genetics laboratories in the future. As extensions of colleges and universities, such laboratories will foster greater interest and more specific training in the field. Students of today, with an eye to the future, may wish to consider entering on the ground floor of what is likely to become an enduring field of research that will profoundly influence our society's educational institutions and health service organizations for many generations to come.

Chapter Summary

Among the outcomes of the Human Genome Project was the identification of 3.7 million SNPs. SNP analysis has the potential to be an effective method of locating subtle genetic variations between individuals that have functional consequences. SNP screening is a potentially useful technique for evaluating susceptibility to a wide range of disorders. The ability to identify such genetic factors before they produce symptoms would greatly benefit applied pharmacogenomics, perhaps most notably in the use of neuroprotectants as a preventive therapy.

The findings of behavior genetics research will directly influence the roles and responsibilities of genetic counselors. As the field of behavior genetics expands, one can envision genetic counselors focusing on specialized subfields in order to keep pace with rapidly accumulating research findings.

Currently, genetic testing is divided into two basic paradigms. Presymptomatic testing is typically conducted in healthy individuals at risk of carrying an allele that is destined to produce a disease state. Susceptibility testing, on the other hand, is used to assess individuals for inheritance of multiple alleles that put them at risk for a disease. Because results from genetic testing sometimes influence the decision to terminate a pregnancy, opponents of abortion may be inclined to oppose genetic testing and behavior genetics research in general. Such opposition would be misdirected, however, considering that a primary goal of behavior genetics research is to establish viable treatments and preventive measures for genetic diseases. Behavior genetics research has also increased public awareness and understanding of genetic disorders, enhancing the re-

catatonic schizophrenia Form of schizophrenia characterized by negative symptoms including prolonged states of stupor and assumption of rigid body positions.

catechol-O-methyltransferase (COMT) gene* Gene that encodes catechol-O-methyltransferase, an enzyme responsible for breaking down and deactivating catecholamine neurotransmitters.

catecholamine Neurotransmitter derived from the amino acid tyrosine (i.e., dopamine, epinephrine, norepinephrine).

centimorgan Unit of measure used to denote a 1% probability that two linked genes will be separated through recombination. One centimorgan is estimated to equal a physical length of approximately 1 million bases.

centromere Region where paired chromosomes are linked together. This region does not contain genetic information.

chorionic villus sampling Procedure in which cells surrounding a developing fetus are removed, typically for testing of fetal DNA for chromosomal abnormalities and genetic diseases.

chromosome An individual body of DNA composed of a long double helix strand.

classic manic depression See *bipolar disorder I*

clozapine Popular atypical neuroleptic used primarily for treatment of schizophrenia.

CLP11 gene* Gene that encodes the P11 protein.

codon Adjoined triplet sequence of nucleotides that codes for a specific amino acid, or initiates or terminates a coding signal.

cognitive ability Tangible, measurable intellectual process such as mathematical ability, reading ability, memory, spatial ability, and production and comprehension of language.

complete AIS Form of androgen insensitivity syndrome in which androgen receptors are completely unresponsive to circulating testosterone. External physical characteristics of an XY individual with complete AIS are nearly indistinguishable from those of an XX individual.

complimentary bases Bases that reliably bond to each other. In DNA, for example, adenine always bonds to thymine and guanine always bonds to cytosine.

concentration gradient Energy force capable of moving ions across the neural membrane, created by an imbalance in ion concentration between the intracellular and extracellular environment. Ions move with a concentration gradient from an area of high concentration to an area of low concentration.

concordance rate Rate at which a trait expressed in one group of individuals is also expressed in a paired group of individuals. Concordance rate is frequently calculated for monozygotic twin pairs or dizygotic twin pairs.

congenital adrenal hyperplasia Disorder in which the adrenal gland releases excess concentrations of androgens during fetal development. In females, this disorder can stimulate male-typical development of bipotential cells and tissue, most notably masculinizing female genital structures. Also known as adrenogenital syndrome.

consanguineous union Marriage between individuals who are blood relatives.

contextual fear conditioning Process in which a subject associates a neutral stimulus or context with a fear-evoking stimulus (e.g., a conditioning cage with an electrical shock). After conditioning has occurred, fear response can be measured by behaviors (e.g., freezing behavior) exhibited in the presence of the neutral stimulus only.

corpus callosum Brain structure that connects the two hemispheres and has a primary function of transmitting neural signals between the hemispheres. The corpus callosum is composed primarily of neuronal axons.

correlational relationship Relationship in which two measures vary together but causation is not established. In a positive correlation both measures move in the same direction (increase/increase or decrease/decrease). In a negative correlation, the measures move in opposite directions (increase/decrease).

cortisol Hormone released by the adrenal gland, frequently in response to psychological or physical stress.

CREB1* gene   Gene that encodes cAMP responsive element–binding protein 1.

criminal behavior   Activity that causes social harm and is punishable by law.

critical period   Range of time in the developmental process during which exposure to certain environmental influences has a relatively greater affect than at other times.

crossing over   Common term for recombination.

cytogenic band   Term used for a chromosome region that stains in a reliable fashion and is used as a landmark for locating genes.

cytokine   Endogenous substance that acts as a chemical messenger for cell-to-cell communication. Some cytokines are released during immune system responses.

cytokine hypothesis of depression   Theory that high concentrations or chronic release of cytokines can produce symptoms of depression by disrupting normal neural signaling.

cytoplasm   Fluid contained within the cell membrane (i.e., intracellular fluid).

cytosine   One of the four organic bases found in DNA and RNA.

D4DR* gene   Gene that encodes the D_4 dopamine receptor protein.

DAT1* gene   Gene that encodes the dopamine transporter protein.

DCDC2* (doublecortin domain containing 2) gene   Gene that encodes a protein with two doublecortin peptide domains.

dementia   Mental deterioration marked by cognitive deficits and memory impairment.

denaturation   Process whereby bonded pairs of DNA strands separate into two individual strands. One of the three basic steps of PCR.

dendrite   Fiberlike projection, usually marked by extensive branching, that extends from the body of a neuron, with a primary function of receiving neural signals.

deoxyribonucleic acid (DNA)   Structure that encodes genetic information; composed of weakly bonded adenine, guanine, cytosine, and thymine molecules.

depolarization   Movement from a state of being electrically charged toward a state of neutral charge. In neurons, depolarization refers to a movement from a negative charge toward a neutral charge in the intracellular environment.

Diagnostic and Statistical Manual of Mental Disorders (*DSM-IV*)   Handbook used for diagnosing mental disorders, published by the American Psychiatric Association.

DiGeorge syndrome   Microdeletion syndrome most commonly caused by a loss of genetic material from region 22q11. Less frequently this syndrome may result from microdeletion on chromosome 10 or 18. Characteristics of this syndrome include developmental delays, learning disabilities, and (in about half of all patients) mental retardation.

disorganized schizophrenia   Type of schizophrenia characterized by inappropriate affect and disorganized thoughts and speech.

dizygotic twins   Twins derived from two separate eggs, each fertilized by a separate sperm cell. These twins share, on average, the same amount of genetic information as do any two nontwin siblings. Also referred to as fraternal twins.

DNA   *See deoxyribonucleic acid*

DNA backbone   Term used to describe the chain of deoxyribose molecules bonded together by phosphate molecules. Four bases are bound to this backbone to complete the structure of DNA.

DNA fingerprint   Unique number and/or distribution of specified DNA sequences used to identify an individual (e.g., for a criminal investigation) or to determine blood relationships (e.g., to establish paternity).

dominance deviation   The relative influence of one allele over a paired allele, or the relative influence of one pair of genes over another pair of genes when both pairs contribute to the same trait.

dominance effect Expression of one pair of genes overshadowing the effect of another pair of genes when both contribute to the same trait.

dominant allele Allele that determines the expressed trait (or has the greatest influence on that trait) when paired with a recessive allele.

dopamine hypothesis of schizophrenia Theory that primary symptoms of schizophrenia are due to excess activity in neural systems that utilize dopamine, most notably the mesolimbic pathway. This theory is based on findings that some dopamine antagonists significantly reduce symptoms of schizophrenia.

Down syndrome Disorder caused by an extra 21st chromosome and characterized by deficits in cognitive ability and abnormalities in some physical features. Sometimes referred to as trisomy 21.

DRD2 gene* Gene that encodes D_2 dopamine receptor proteins.

Duchenne muscular dystrophy Most common form of muscular dystrophy diseases that are characterized by progressive muscle weakness and degeneration of muscle tissue.

dyscalculia Learning disability characterized by impairments in using or comprehending symbolic representations in mathematical calculations.

dysgraphia Learning disability characterized by impairments in ability to write or produce legible handwriting, regardless of ability to read and produce spoken language.

dyslexia Learning disability characterized by impairments in reading and producing written language.

dysthymia Chronic state of mild depression.

dystrobrevin-binding protein 1 (DTNBP1) gene* Gene that encodes dystrobrevin-binding protein.

dystrophy disease Type of genetic disease that results in poorly developed or rapidly degenerating tissues (e.g., muscular dystrophy results in muscle tissue atrophy).

early-onset Alzheimer disease Form of Alzheimer disease that manifests before the age of 65. This form of Alzheimer is highly heritable, with some family studies showing a dominant single-gene inheritance pattern.

Edward syndrome Trisomy disorder, caused by the addition of an extra 18th chromosome, that is usually fatal. In individuals who survive beyond one year, marked physical and behavioral defects are noted, including mental retardation.

Ego Dystonic Homosexuality (EDH) Mental illness category added to the *DSM-III* in 1980 to replace homosexuality as a diagnosable mental illness. Criteria for EDH were: 1) persistent lack of heterosexual arousal, which the patient experienced as interfering with initiation or maintenance of wanted heterosexual relationships, and (2) persistent distress from a sustained pattern of unwanted homosexual arousal. EDH was removed without replacement in 1987 in the DSM-III-R.

EKN1 gene* Candidate gene for dyslexia whose function is currently unknown.

electrical gradient Energy force capable of moving ions across the neural membrane, created by an imbalance in electrical charge between the intracellular and extracellular environment. Ions always move with an electrical gradient from an area sharing their electrical charge to an area with an opposite charge (i.e., positive ions will move from a positively charged environment to a more negatively charged environment).

electroencephalography (EEG) Recording of gross electrical activity of the brain using electrodes placed on the scalp.

electrophoresis Process of drawing stained segments of DNA through an agarose gel block using an electrical charge. Segment size can be estimated by the relative location of stained bands in the gel block when the process is completed.

elevated plus maze Cross-shaped maze with two open arms and two closed arms, elevated at least 1 meter off the floor. This maze is used to assess fear/anxiety in rodents.

endogenous reward system   Theoretical neural system, believed to be integrated with the mesolimbic dopamine system, that is activated when an animal engages in beneficial behaviors (e.g., eating, drinking, sex behavior).

enriched environment   Environment that contains a variety of stimuli capable of enhancing neural development (e.g., varied sounds, visual cues, textures).

environmental alteration   Change made to typical environmental experiences in an effort to alter expression or development of a phenotype (e.g., decreasing phenylalanine ingestion in an effort to reduce effects of PKU).

environmental factor   Factor other than genes that influences phenotype.

enzyme   Protein with a primary function of facilitating chemical reactions or metabolic processes.

epistasis   When expression of one genetic trait masks the effects of a second genetic trait (e.g., genes for premature baldness may mask genes for premature gray hair).

estrogen receptor beta (*ERβ*) gene*   Gene that encodes the beta subtype of the estrogen receptor protein.

eugenics   Theory that both physical and mental traits could be improved in the human race through selective breeding. First formally proposed in 1869 by Sir Francis Galton.

eukaryotic cell   Cell containing a nucleus.

evocative trait   Genetically influenced trait that tends to evoke a response from others (e.g., extroversion trait may increase the number of people who choose to interact socially with an individual).

evolution   Change in allele frequencies in a population that ultimately increases the diversity and adaptation capacity of those organisms.

exon   DNA sequence that leaves the nucleus and codes for production of a protein.

expanded triplet repeat disorder   Genetic disorder caused by the expansion of a repeated three-base coding sequence from a normal range to an abnormally high range. Examples include Huntington disease and fragile X syndrome.

extension   Process in which DNA primers initiate development of a complementary strand to a single strand of DNA. One of the three basic steps of PCR.

factorial analysis   Statistical method used to determine the influence of multiple factors (e.g., memory, verbal ability, mathematical ability) on variability in an observed measure (e.g., intelligence test score).

familial Alzheimer disease   Form of early-onset Alzheimer disease that follows a dominant single-gene inheritance pattern.

familial DiGeorge syndrome   Rare form of DiGeorge syndrome that may be noted with relatively high frequency within a family pedigree.

familial pattern   Inheritance pattern that can be identified in members of the same family, usually over multiple generations.

family history   Information regarding the presence of traits or diseases in family members, usually spanning multiple generations.

family-based genetic association study   Study designed to identify genetic markers shared by related individuals who share a phenotype of interest.

FAS   See *fetal alcohol syndrome*

FAT* gene   Gene that encodes a very large protein of the cadherin family.

fear conditioning   Technique in which a neutral stimulus is paired with a noxious stimulus, resulting in an association such that the neutral stimulus evokes a fear response.

fetal alcohol syndrome (FAS)   Syndrome caused by exposure to alcohol during a critical period of fetal development. Features include small head, delays in learning, cognitive impairment, and emotional problems.

first-degree relatives   Term used for relatives who share 50% (or an average of 50%) of their genetic material with an individual; these include the individual's parents, siblings, and children.

Five Factor Model (FFM)   Self-rating scale that assesses personality using questions based on the Big Five dimensions of personality.

five prime (5′)   Designation given to the end of a nucleotide strand that has a free (unbound) carbon atom in the 5′ position on the sugar molecule.

flanking region   Two end regions (beginning at 5′ and 3′) of a segment of DNA being analyzed by PCR.

fragile X mental retardation 1 (*FMR1*) gene*   Gene that encodes an RNA-binding protein.

fragile X syndrome   Disorder caused by a large triplet repeat expansion within the *FMR1* gene. Fragile X syndrome is a leading cause of mental retardation and developmental disabilities in males.

frontal lobe   Anterior (frontal) region of the cerebral cortex primarily responsible for controlling higher level cognitive function, some aspects of voluntary movement, and speech production.

frontotemporal dementia   Form of early-onset dementia with the greatest level of dysfunction seen in frontal and temporal lobe functions. Pick disease is the most common form of this type of dementia.

functional genomics   Field of research that utilizes findings from genomic projects to describe functions and interactions of both genes and the proteins whose construction they direct.

G8 marker   Readily identified DNA sequence on chromosome 4 that is closely linked to the Huntington allele.

gamete cell   Reproductive cells (i.e., sperm, egg, pollen) that contain one half the compliment of DNA found in nongamete cells.

gamma-aminobutyric acid (GABA)   Neurotransmitter that acts on chloride channels to reduce firing of cells.

gene   Fundamental physical unit of heredity composed of a particular segment of DNA.

gene conversion   Process whereby one chromosome physically incorporates a segment of DNA from the same region of a homologous chromosome. Unlike recombination, which is a reciprocal exchange of genetic material between homologous chromosomes, gene conversion is nonreciprocal.

gene therapy   Treatment of diseases or disorders through alteration of gene function. Gene therapy is an applied subfield of genetic engineering.

gene–environment correlation   When an individual with a genetic predisposition for a particular trait is also exposed to environmental factors that foster that same trait.

gene–environment interaction   When gene expression is directly altered by environmental influences.

general intelligence factor (*g* factor) (*g*)   Theoretical common element that contributes to the results of all tests of intelligence.

genetic anticipation   Phenomenon in which successive generations experience an earlier onset, or a more severe form, of a genetic disease than previous generations.

genetic diversity   Reference to the range of genetic characteristics and the genetic composition (allelic variation) within a species. Greater genetic diversity is generally associated with greater adaptability of a species.

genetic engineering   Process of inserting new genetic information into a cell with the intention of changing that cell's functions and ultimately modifying the phenotype of the organism.

genetic marker   Readily identifiable feature of DNA that is consistently found in the same location on a given chromosome. Genetic markers are often used to locate specific genes and, in some cases, may be located within a gene sequence.

genetic predisposition Term used to describe a gene or combination of genes that produce specific traits only when the organism is exposed to specific environmental factors.

genetically modified stem cell lines Stem cells altered through genetic engineering to increase viability (e.g., increase replication capability, increase resistance to immunorejection).

genetics Study of genes and heredity.

genome Entire compliment of nuclear DNA in a given organism.

genomic imprinting Phenomenon whereby the sex of the parent contributing the gene defect significantly affects expression of the associated trait. See *Angelman syndrome* and *Prader-Willi syndrome* for examples.

genomics Field of study concerned with genes and gene functions.

genotype Specific gene or genes responsible for a phenotype.

germ cell Reproductive cells (i.e., sperm, egg, pollen) that contain one half the compliment of DNA found in nongamete cells.

germ cell gene therapy Gene therapy directed at altering germ (gamete) cells with the goal of inducing changes that are transmissible over generations.

germinal mosaicism Condition where a genetic defect is carried in one parent's gamete cells but is not found in nongamete cells.

GLO1 gene* Gene that encodes the enzyme glyoxalase I.

glucocorticoid Steroid hormone produced and released by the adrenal gland.

glutamate Excitatory amino acid utilized as a neurotransmitter.

glutathione reductase 1 Enzyme believed to be essential to antioxidant defense mechanisms.

glyoxalase 1 Enzyme believed to have functions similar to those of glutathione reductase 1

Golgi apparatus Intracellular organelle responsible for processing and packaging newly formed proteins.

GSR gene* Gene that encodes the enzyme glutathione reductase 1.

guanine One of the four organic bases found in DNA and RNA.

health psychology Field psychology concerned with the impact of psychological function on physical well-being and illness.

hemophilia Disorder characterized by a deficiency in blood clotting; the most common type is caused by a recessive single-gene aberration on the X chromosome.

heritability Extent of genetic influence on trait variation and, subsequently, the probability that a particular trait variant will be passed on to successive generations.

heteroplasmy Presence of more than one type of mitochondrial DNA in an single individual.

heterozygote Individual who carries two different variations (alleles) of the same gene on a chromosome pair.

homozygote Individual who carries two of the same gene type (allele) on a chromosome pair.

hSERT gene* Gene that encodes a serotonin transporter protein.

human genome Entire compliment of nuclear DNA in a human. Current estimates are between 20,000 and 25,000 genes.

Human Genome Project International research effort initiated in 1990 with a primary goal of identifying and sequencing the entire human genome.

huntingtin Protein whose production is controlled by the Huntingtin gene. Function of this protein is currently unknown.

Huntingtin (HD) gene* Gene that encodes the protein huntingtin.

Huntington disease Autosomal dominantly inherited disease that results from expansion of a triplet repeat within the huntingtin gene. Primary characteristics include progressive motor disturbances that typically begin in midlife.

hydrophilic Property of being attracted to water.

hydrophobic Property of being repelled by water.

hypertension Chronic elevated blood pressure.

hypochondria Chronic preoccupation with, and anxiety related to, an imagined illness.

hypophosphatemia Condition characterized by abnormally low concentrations of phosphate in the blood.

hypothalamic–pituitary–adrenal (HPA) axis Reference used to describe the interrelated hormone feedback loop between these three glands. HPA activity increases in times of psychological arousal (particularly stress).

idiopathic Descriptor used for a phenomenon with an unknown cause (e.g., idiopathic disease).

impoverished environment Environment lacking adequate stimuli for normal neural and behavioral development.

in utero gene therapy Alteration of gene function in a developing fetus with the objective of treating a genetic disease or disorder. In utero gene therapy is an applied subfield of genetic engineering.

inbreeder Term used to describe plants capable of fertilizing themselves.

induced mutation Genetic mutations that occur as a result of exposure to exogenous factors such as chemicals, drugs, radiation or viruses.

inhibition Term used to describe the process of reducing the likelihood that a neuron will transmit a signal.

intelligence An individual's abilities to reason, solve problems, comprehend ideas, and learn.

intelligence quotient (IQ) test Test designed to assess a variety of cognitive abilities in an effort to quantify general intellectual ability.

intersexual selection Process of mate selection that is influenced by the relative "attractiveness" of an individual to members of the opposite sex.

intragender sexual action Term sometimes used to describe homosexual behavior among nonhuman species.

intrasexual selection Process of mate selection that is influenced by an individual's ability to compete with others of the same sex for access to mates.

intron DNA sequence within a gene that does not leave the nucleus and has no apparent role in coding for production of a protein.

ion channels Proteins imbedded in neuronal membranes that allow the flow of ions into and out of the cell.

ionotropic receptor Receptor sites that are directly linked to ion channels. Activation of this type of receptor results in a rapid change in membrane potential, producing either excitation or inhibition of a cell.

IQ QTL Project Project undertaken to assess and compare frequency of particular alleles in individuals who rank in the high and low extremes of the IQ scores.

junk DNA Term sometimes used to describe DNA found outside of gene regions. Junk DNA constitutes over 95% of total DNA. This term may be misleading, since this large percentage of DNA may have some undiscovered role in genetic transmission or function.

karyotype Photograph of all the chromosomes, arranged in homologous pairs, from a given individual.

KIAA0319 gene* Candidate gene for dyslexia that is known to be involved in neuronal migration.

Klinefelter syndrome Trisomy syndrome with an XXY genotype, characterized by a male phenotype with feminized features.

knock-in animal Transgenic animal in which a segment of DNA has been introduced into the normal genome using genetic engineering techniques.

knock-out animal Transgenic animal in which a segment of DNA has been removed from the normal genome or inactiviated/disrupted using genetic engineering techniques.

learning disability Deficit in understanding or using spoken or written language, usually characterized by a restricted capacity to concentrate, listen, speak, read, write, or use mathematical calculations.

Lewy bodies Abnormal, round, protein-based structures found in the cytoplasm of brains from individuals with Parkinson disease and some other forms of dementia.

Lewy body dementia Form of dementia with characteristics of both Alzheimer disease and Parkinson disease (including the presence of Lewy bodies).

liability threshold Concept that as an increasing number of alleles that contribute to a given polygenic trait are inherited, an individual nears a threshold for trait expression. Once the threshold is reached, the likelihood that the trait will be expressed increases dramatically.

light–dark box test Testing paradigm in which rodents are given a choice between a light (open) arena and a dark (enclosed) arena. Excessive time spent in the dark arena is associated with higher levels of timidity, anxiety, or fear.

linkage Instance where two genes are located in close proximity on the same chromosome, resulting in frequent coexpression of independent genetic traits.

linkage analysis Process of determining the relative location of two or more genes on the same chromosome by establishing how frequently they are coexpressed (i.e., how closely they are linked).

lipid bilayer Arrangement of lipid molecules into two adjacent rows making up the primary component of a neural membrane.

lithium Chemical that is used in a carbonate form to stabilize mood in individuals with bipolar disorder.

locus (pl., loci) Position on a chromosome of a gene or gene marker.

major depression Mood disorder characterized by chronic feelings of dysphoria, worthlessness, helplessness, sadness, despair, and lethargy.

male-specific region Portion of the Y chromosome that does not have homologous regions on the X chromosome.

manic depression See *bipolar disorder*

MeCP2 gene* Gene that encodes methyl-CpG-binding protein 2 (MeCP2), which is believed to inhibit gene transcription.

medial preoptic area (MPOA) Sexually dimorphic brain structure located adjacent to the hypothalamus that is larger in men than in women.

meiosis Process of cell division used by gametes, whereby division of a parent cell results in daughter cells that each contain half the original number of chromosomes.

membrane Outer portion of a cell that separates the intracellular environment from extracellular fluid.

Mendel's law of independent assortment
Theory that the probability of inheriting any given gene is unaffected by the probability of inheriting a second, separate gene.

Mendel's law of segregation Theory that during the process of meiosis paired alleles randomly separate into gametes that in turn carry only one allele for any given trait.

mental retardation Condition of low general intellectual function that includes impaired ability to acquire skills required for normal daily activities.

mesolimbic dopamine system Primary dopamine pathway in the brain that is largely associated with mood and emotion.

messenger RNA (mRNA)   RNA that carries a template for protein synthesis from the nucleus to the ribosomes.

metabotropic receptor   Protein imbedded in the neural membrane that, when chemically activated, initiates a cascade of events ultimately increasing or decreasing cell activity, usually by altering ion flow through nearby membrane protein channels.

methylation   Addition of a methyl group. When methylated during early development, cytosine is rendered inactive, inhibiting genetic effects.

methyl-CpG-binding protein 2   Regulatory protein that directs production of several other proteins. Mutations of a gene that produces this protein are associated with Rett syndrome.

microdeletion syndrome   Syndrome caused by a chromosomal deletion too small to be detected visually in a karyotype.

microtubule-associated protein tau (MAPT) gene*   Gene that encodes the microtubule-associated protein tau and has been associated with development of Alzheimer disease.

mild mental retardation   Generally defined as having an IQ score between 55–69, with an ability to learn reading, writing, and mathematic skills at a grade school level.

mirror twins   Monozygotic twins with distinctive features on the left side of one sibling appearing on the right side in the other, and vice versa. Mirror twins result from late splitting of the embryo 8–12 days after fertilization.

mitochondria   Intracellular organelles that convert energy stores to a form of energy called ATP that can then be used for most cellular functions.

mitosis   Process of a nongamete cell replicating to form two daughter cells with identical sets of chromosomes.

moderate mental retardation   Generally defined as having an IQ score between 40–54 with limited reading, writing, and math skills. Noticeable motor skill impairment and speech deficits are also typical.

monosomy   Condition of a having a only one copy of a chromosome

monozygotic twins   Twins derived from a single egg cell that is fertilized by a single sperm cell.

Morris water maze   Cognitive testing paradigm developed by Richard Morris, in which a rodent, placed in a large tub of water, searches for a platform hidden below the water's surface. This paradigm tests several cognitive skills, most notably spatial memory.

multiple sclerosis (MS)   Chronic genetic disease characterized by progressive loss of myelin, leading most notably to decreased muscle control.

mutagen   Substance or condition that causes a permanent change in genetic structure.

mutant   Organism that results from a genetic mutation.

mutation   Change in the number, arrangement, or molecular structure of one or more genes.

natural selection   Process whereby some individuals survive longer and reproduce more than their conspecifics, thereby increasing the frequency of their alleles in subsequent generations.

nature-versus-nurture   Topic of debate concerning the relative importance of genes (nature) and environment (nurture) on behavioral traits.

negative symptoms   When describing features of schizophrenia, refers to a lack of emotional expression and reduced motor activity (e.g., catatonia).

neural Darwinism   Term sometimes used to describe the process of apoptosis as it applies to pruning of nonessential neural connections during development.

neurofibromatosis type 1 (NF1)   Single-gene dominant disorder that produces glial cell tumors and brown spots on the skin.

neurogenetics   Field of research interested in understanding the genetic basis of neural structure and function.

neuroleptic Class of dopamine antagonists used primarily in the treatment of schizophrenia.

neurological realm Subfield of gene therapy research concerned with developing treatments to reduce cell loss caused by acute and chronic neurological disorders such as stroke and Parkinson disease, respectively.

neuromodulator Substance, usually released by neurons, that acts to increase or decrease the effect of neurotransmitters.

neuropharmacology Field of research interested in understanding how drugs influence neural function.

neuroprotectant Chemical compound capable of slowing or halting cell death associated with a neurodegenerative process.

neuroregulin 1 (*NRG1*) gene* Candidate gene for schizophrenia that encodes the protein neuroregulin 1.

neurotransmitter Endogenous substance that alters cell activity, typically by binding to protein receptor sites.

neurotransmitter receptor site Membrane-embedded protein that alters neural activity when it is bound by a neurotransmitter.

neurotransmitter reuptake Process whereby neurotransmitter molecules released into the synapse are recaptured by a terminal for reuse (a form of neurotransmitter recycling).

***N*-methyl-D-aspartate (NMDA)** Amino acid derivative that acts as an agonist at the NMDA receptor subtype of the glutamate receptor protein.

nondisjunction Genetic error occurring during cell division where a chromosome pair fails to properly separate into daughter cells.

nonpurging bulimia Bulimia characterized by episodes of binging followed by extended periods of abstinence from eating.

nonshared environment Any environmental factors that differ between individuals (e.g., siblings) that have the potential to interact with genetic influences on behavior.

norepinephrine transporter (*NET*) gene* Gene that encodes the norepinephrine transporter protein.

normal distribution Theoretical data distribution represented by a bell-shaped curve indicating the highest frequency occurring in the midrange of the distribution with approximately equal low frequency occurring at the tails of the curve.

normal range Term used to describe scores (e.g., IQ) that do not differ significantly from the population mean.

nuclear RNA (nRNA) RNA found within the nucleus of the cell.

nucleotide Name given to a DNA component composed of three molecules (sugar, base, and phosphate).

nucleus Central component of a cell that contains RNA and cellular DNA.

obesity Condition of being overweight, typically defined by a weight of greater than 20% above that recommended for an individual's height, sex, and age.

obsessive–compulsive disorder Anxiety disorder characterized by persistent thoughts (obsessions) or behaviors (compulsions) that produce anxiety severe enough to disrupt normal daily activities.

open field activity Measure of locomotion and exploratory behavior generated by assessing activity of a rodent in an open chamber.

opioid neuropeptide Endogenous chemical substance that acts as a neurotransmitter on opioid receptor sites.

OPRM1* gene Gene that encodes the mu opioid receptor protein.

oxidative phosphorylation Enzyme-mediated metabolic process that synthesizes ATP required for cellular energy.

oxidative stress Condition associated with a variety of diseases that results in free radical production believed to stimulate neurodegeneration through a process of apoptosis.

p   Used to denote the short arm of a chromosome.

p11 protein   Protein believed to be involved in mood regulation and implicated in the development of depression.

palindrome   Sequence that reads the same when read forward or backward.

Parkinson disease   Neurodegenerative disease of aging characterized by a loss of dopamine-producing cells in the substantia nigra, slowed movement, and decreased emotional affect.

parkinsonian akinesia   Absence of movement caused by Parkinson disease or by drugs that adversely inhibit nigrostriatal dopamine transmission (e.g., some neuroleptics).

parthenogenic activation   Rare process of egg cell division that begins in the absence of fertilization.

partial AIS   Form of AIS in which androgen receptor proteins are only partially responsive to androgen. XY individuals with partial AIS may develop genitalia that are intermediate between male and female.

passive correlation   When alleles that evoke a trait correlate by chance with environmental conditions thought to evoke the same trait.

Patau syndrome   Typically fatal trisomy condition characterized by the addition of a 13th chromosome; those who do survive show moderate to severe mental retardation.

PCR   See *polymerase chain reaction*

pedigree chart   Chart showing a family history of some inherited trait(s).

personality disorder   Persistent nonpsychotic state of functional impairment and/or distress caused by the use of maladaptive and inflexible behavior patterns to attain self-satisfaction.

pharmacogenomics   Field of research interested in understanding the influence of genes on an individual's response to drug treatment.

phenotype   Expressed characteristics of a genotype (e.g., physical characteristics, behavior, personality, illness).

phenylketonuria (PKU)   Autosomal recessive disorder characterized by an inability to metabolize phenylalanine, ultimately producing neurotoxic effects marked most notably by developmental delays and mental retardation.

Pick disease   Common form of frontotemporal dementia, usually diagnosed between the ages of 40 and 60.

PKU   See *phenylketonuria*

polygenic trait   Trait mediated by multiple (usually many) genes.

polymerase   Enzyme that catalyzes DNA and RNA construction and a primary component in the polymerase chain reaction.

polymerase chain reaction (PCR)   Method of replicating DNA that is capable of creating millions of segment copies within several hours.

polymorphism   DNA sequence variation that may have functional consequences.

population-based genetic association study   Study designed to identify genetic markers shared by unrelated individuals who exhibit a phenotype of interest.

positive symptom   When describing features of schizophrenia, refers to delusions, hallucinations, incoherent speech, and abnormal motor activity.

postnatal environmental effect   Environmental factors encountered after birth that are capable of altering phenotype expression.

postsynaptic membrane   Membrane of a cell that receives information from another cell; usually contains some form of receptor protein.

Prader-Willi syndrome   Genetic disorder caused when a deletion from 15q11-q13 is inherited from the father. Symptoms include some language delays, obsession with food, and obesity. See also *genomic imprinting* and *Angelman syndrome*.

permutation   Unstable mutation that does not produce phenotypic effect but may expand in successive generations to a full mutation that produces effects (e.g., fragile X syndrome has been identified in a premutation form).

prenatal environmental effects Environmental factors encountered in utero that are capable of altering phenotype expression.

presenilin 1 Protein found in the brain and believed to be involved with neuronal signaling and positioning of other proteins within the neuron.

presenilin 1 (*PSEN1*) gene* Gene that encodes the protein presenilin 1.

presenilin 2 Protein found in the brain and believed to function similarly to presinilin 1.

presenilin 2 (*PSEN2*) gene* Gene that encodes the protein presenilin 2.

presymptomatic testing Testing of healthy individuals for genetic abnormalities known to produce diseases (e.g., Huntington disease, PKU).

presynaptic membrane Membrane of a cell that sends information to another cell, usually by releasing some form of transmitter substance (e.g., hormone, neurotransmitter).

primer Short base sequence (typically less than 30 bases) that initiates DNA replication required for PCR.

proband Starting point, or index case, for a pedigree chart.

profound mental retardation Generally defined as having an IQ score of less than 25. Comorbid medical problems and severe communication and motor impairments are also common.

proinflammatory cytokine Class of cytokines released during acute illness that promote fever, inflammation, and tissue destruction.

prokaryote Organism that lacks a nucleus.

protein Large molecule composed of amino acids.

proteome All of the proteins produced by the genes of a given species.

proteomics Study of proteins and proteomes.

psychiatric realm subfield of gene therapy research concerned with developing treatments for neural dysfunctions associated with psychiatric disorders such as schizophrenia, depression, etc.

psychometric theory Theory that intelligence is a combination of abilities that can be measured by testing multiple cognitive functions.

psychoneuroendocrinoimmunology (PNEI) Recently developed field of research that assesses the interactions between psychology, the nervous system, the endocrine system, and the immune system.

psychophysiological illness Recently introduced term given to a physical illness that is caused, or exacerbated, by psychological factors. This term is preferred by some to the older term of psychosomatic disorder.

psychosomatic disorder See *psychophysiological illness*

ptel Term for the telomere region of the short arm of a chromosome.

punctuated equilibrium A theory of evolution that suggests long periods of limited change are punctuated by short periods of extensive change, including speciation.

Punnett square A grid-based graphical illustration developed by Reginald Punnett to illustrate offspring genotype of two parents.

purging bulimia Eating disorder characterized by bouts of excessive caloric intake followed by self-induced vomiting or use of laxatives.

purine Two-ring molecule composed of four nitrogen atoms bonded with carbons.

pyrimidine One-ring molecule composed of two nitrogen atoms bonded with carbons.

q Used to denote the long arm of a chromosome.

qtel Term for the telomere region of the long arm of a chromosome.

qualitative trait Trait usually regulated by a single gene that varies in a discrete rather than continuous manner.

quantitative trait   Trait usually regulated by multiple genes that varies in a continuous rather than discrete manner.

quantitative trait loci (QTL)   Genes that contribute in varying degrees to a common trait that is expressed quantitatively.

Raven's progressive matrices   Multiple-choice test used to measure cognitive abilities independent of language skills by allowing subjects to correctly complete a pictorial sequence.

reactive correlation   When alleles evoke expression of a trait which, in turn, evokes responses from the environment that encourage the same trait.

reception   In a neuron, the process of receiving a signal produced by the environment or from another neuron.

receptor   Protein capable of altering cell function by binding and responding to chemical substances (e.g., neurotransmitters, hormones, drugs).

recessive allele   An allele that must be present on both paired chromosomes to be fully expressed in the phenotype.

recombination   A reciprocal exchange of genetic information between two homologous chromosomes during gamete formation.

resting potential   The difference in electrical charge between intracellular and extracellular fluid for a neuron at rest (typically about –70 mV).

restriction enzyme   An enzyme that recognizes specific DNA sequences and cleaves the DNA strands at those regions.

retrovirus   Type of virus that uses reverse transcriptase to insert its genetic material into the DNA of infected cells. This type of virus is among the most common vectors used for genetic engineering and gene therapy research.

Rett syndrome   X-linked genetic disorder that is fatal in males, and produces mental retardation, seizures, and other neurological problems in females.

reverse transcription   Synthesis of a segment of DNA from a strand of RNA promoted by the enzyme reverse transcriptase.

ribonucleic acid (RNA)   A nucleic acid molecule similar to DNA but containing ribose rather than deoxyribose and consisting of weakly bonded adenine, guanine, cytosine, and uracil molecules.

ribosome   An intracellular organelle responsible for synthesizing proteins from amino acids.

RNA   See *ribonucleic acid*

RNA splicing   The process of removing introns from a primary transcript and combining exons into a functional mRNA coding sequence.

RNA-induced silencing complex (RISC)   cellular protein that unwinds siRNA and guides it to the appropriate mRNA sequence.

ROBO1* gene   Gene that encodes a protein believed to be involved in guiding axons during development.

rough endoplasmic reticulum (RER)   Intracellular membranous tubes that contain ribosomes.

schizophrenia   Severe chronic psychopathology characterized by inappropriate emotions or affect, disconnected or illogical thinking, and socially inappropriate behaviors.

schizophrenia-related personality disorders   Collective term used by some researchers for the following personality disorders: paranoid, schizoid, schizotypal, borderline, and avoidant.

schizophrenia-spectrum disorders   Disorders identified in the pedigree of a schizophrenia patient that suggest the presence of alleles that may contribute to schizophrenia (e.g., schizophrenia-related personality disorders).

SCID   See *severe combined immune deficiency*

second messenger   Molecules synthesized within the cell in response to receptor protein activation. Second messengers initiate a cascade of intracellular changes that subsequently alter cell metabolism or activity.

second-degree relatives   Term used for relatives who share 25% (or an average of 25%) of their genetic material with an individual; these include the individual's grandparents, grandchildren, siblings of parents, and children of siblings.

selective serotonin reuptake inhibitor (SSRI)   Class of drugs designed to impair serotonin reuptake proteins, thereby increasing availability of serotonin in the synapse.

self-replication   Process whereby paired DNA strands unbond and then bond free-floating molecules to create two separate and identical paired strands.

severe combined immune deficiency (SCID)   Rare genetic disorder characterized by immune system deficiencies that typically results in early death.

severe mental retardation   Generally defined as having an IQ of 25–39 and characterized by limited communication and motor skills.

sex chromosome   Chromosome designated by X or Y that directly influences development of primary sex characteristics.

sexual selection   Theory that traits providing an advantage in reproductive success over conspecifics of the same sex will be passed on to future generations.

sexually dimorphic   Having a different morphology depending on the sex, as with the medial preoptic area that is larger in males than in females.

shared environment   Any environmental factors shared by individuals (e.g., siblings) that have the potential to interact with genetic influences on behavior.

single nucleotide polymorphism (SNP)   Variation in DNA sequence in which a single base differs from the base that is usually located at that position.

situs inversus   Condition in which major organs form in a mirror position compared to normal development (e.g., heart on the right side of the chest cavity), sometimes seen in mirror twins.

Slit protein   Family of proteins believed to play a role in guiding neural migration, promoting neural outgrowth, and regulating the formation of connections among neurons.

SLITRK1 gene*   Gene that encodes production of the SLITRK1 protein (from the family of Slit proteins).

small interfering RNA (siRNA)   Short artificial segments of double-stranded RNA designed to induce release of "slicer" enzymes capable of cleaving mRNA into nonfunctional segments.

snake-like view   Method of illustrating amino acid chain sequences used in the construction of specific proteins.

solute carrier (SLC6A3) gene*   Gene that encodes a dopamine transporter protein.

soma   Body of a neuron and location of the nucleus.

somatic cell   Nongamete cell.

speciation   Evolutionary process whereby natural selection leads to development of a new species.

spontaneous mutation   A change in the normal DNA structure within a gene, that occurs in the absence of an obvious external influence.

sporadic Alzheimer disease   Occurrence of a case of Alzheimer disease in the absence of any family history of that disease.

sporadic case   Occurrence of a trait in the absence of any family history of that trait.

SSRI   See *selective serotonin reuptake inhibitor*

start codon   Series of three bases used to initiate gene translation. Most frequently AUG, but sometimes CUG and UUG.

stem cell   Immature, undifferentiated cell from which numerous specific cell types may develop.

stop codon   Series of three bases used to terminate gene translation. Most frequently UAA, UAG, or UGA.

sub-band   Term used for a chromosome region that is lightly stained and located between more darkly stained regions known as "bands."

Sub-bands are used as landmarks for determining the location of genes.

substantia nigra   Neural structure containing an abundance of dopamine-releasing neurons that control movement. This structure degenerates in Parkinson disease.

susceptibility gene   Gene for which particular alleles are known to increase the likelihood of expressing a certain trait, most notably development of a disease.

susceptibility testing   Assessing an individual for multiple alleles that put them at risk for a disease (e.g., late-onset Alzheimer disease, schizophrenia).

synapse   The space between a presynaptic membrane where transmitter chemicals are released and a postsynaptic membrane that contains transmitter chemical receptor sites.

tau   Protein involved in intracellular transport of chemicals (e.g., neurotransmitters).

telomere   End region of a chromosome.

temperament   Innate and enduring behavioral trait.

teratogen   Agent or substance that may cause developmental defects when exposure occurs during fetal development.

third-degree relative   Term used for relatives who share 12.5% (or an average of 12.5%) of their genetic material with an individual; these include the individual's great grandparents, great grandchildren, first cousins, and grandchildren of siblings.

three prime (3')   Designation given to the end of a nucleotide strand that has a free (unbound) carbon atom in the 3' position on the sugar molecule.

threshold level   Point at which a rise in intracellular electrical potential triggers the opening of voltage-sensitive protein channels (typically about –50 mV for sodium channels).

thymine   One of the four organic bases found in DNA.

Tourette syndrome   Neurological disorder characterized by uncontrollable tics and, in some cases, vocalizations. Tourette syndrome is sometimes comorbid with symptoms of obsessive–compulsive disorder.

trait breeding   Breeding animals with a shared trait in an effort to produce offspring who also exhibit that trait.

transcription   The construction of a messenger RNA molecule created from a segment of DNA that is used as the template.

transduction   The transformation of a neural signal (sensory or chemical) into cellular energy (membrane potential change).

transfer RNA (tRNA)   RNA that carries amino acids to ribosomes for use in protein production.

transgenerational transmission   Passing of an environmentally induced mutation or pathogen from parent to offspring.

transgenic   Having a genome that has been artificially altered such as a knock-in or knock-out animal.

transient ischemic attack   Temporary interruption of blood flow to a part of the brain, usually lasting only a few minutes, that results in symptoms of a mild stroke; sometimes referred to as a "mini-stroke."

translation   RNA-controlled process of assembling amino acids into proteins.

translocation   Exchange of genetic material between nonhomologous chromosomes.

transmission   Sending a message, as with the release of neurotransmitter.

trichotillomania   Disorder characterized by compulsive hair pulling.

triplet repeat expansion disorder   Class of disorders that includes Huntington disease and fragile X syndrome, where a specific three-base sequence repeat within a gene is expanded to a length that produces a dysfunctional phenotype.

trisomy 21   See *Down syndrome*

trisomy X (XXX)   Addition of a 3rd X chromosome to a female genotype. Typical phenotype is characterized by a taller-than-average female with some developmental delays.

true breeder   Self-fertilizing plant in which traits of offspring always resemble those of the parent (unless the parent is crossbred or artificially fertilized).

Turner syndrome   Monosomy syndrome with an X0 genotype, characterized by a female phenotype that is typically sterile, short in stature, and exhibits some developmental delays and minor physical malformations.

type A   Theoretical personality type marked by aggression, impatience, a sense of time urgency, and a desire to achieve recognition and advancement.

typical neuroleptic   Early class of drugs used to treat symptoms of schizophrenia by blocking dopamine receptors, particularly the D2 receptor subtype.

unbalanced translocation   Phenomenon in which one chromosome loses genetic material that is integrated into a nonhomologous chromosomes in a process of dislocation and fusion.

unipolar mania   Occurrence of mania symptoms in the absence of depression symptoms. The *DSM-IV* does not recognize unipolar mania as a diagnosable mood disorder.

uracil   One of the four organic bases found in RNA.

vascular dementia   Dementia produced by interruption of blood flow to the brain.

vector   Agent used as a carrier for delivery of substances into an organism. Vectors are used in genetic engineering to deliver modified DNA.

vesicle   Spherical membrane that contains chemicals for neural communication (e.g., neurotransmitter), usually located in the axon terminal region.

voltage-sensitive   A type of ion channel that opens and closes in response to electrical potential changes across the cell membrane.

XXXX syndrome   Addition of two X chromosomes to a female genotype. Typical phenotype includes features similar to (and sometimes more profound than) those seen in trisomy X.

XXYY syndrome   Addition of an X and a Y chromosome to a male genotype. Typical phenotype includes features similar to (and sometimes more profound than) those seen in Klinefelter syndrome.

PHOTO CREDITS

REFERENCES

Aalfs, C. M., Smets, E. M. and Leschot, N. J. (2007). Genetic counselling for familial conditions during pregnancy: A review of the literature published during the years 1989–2004. *Community Genetics, 10*(3), 159–168.

Abelson, J. F., Kwan, K. Y., O'Roak, B. J., Baek, D. Y. et al. (2005). Sequence variants in SLITRK1 are associated with Tourette's syndrome. *Science, 310*(5746), 317–320.

Abou Jamra, R., Becker, T., Georgi, A., Feulner, T. et al. (2008). Genetic variation of the FAT gene at 4q35 is associated with bipolar affective disorder. *Molecular Psychiatry, 13*(3), 277–284.

Adams, L. J., Mitchell, P. B., Fielder, S. L., Rosso, A. et al. (1998). A susceptibility locus for bipolar affective disorder on chromosome 4q35. *American Journal of Human Genetics, 62*(5), 1084–1091.

Aghajanian, G. K. and Marek, G. J. (2000). Serotonin model of schizophrenia: Emerging role of glutamate mechanisms. *Brain Research: Brain Research Reviews, 31*(2–3), 302–312.

Airaksinen, E. M., Matilainen, R., Mononen, T., Mustonen, K. et al. (2000). A population-based study on epilepsy in mentally retarded children. *Epilepsia, 41*(9), 1214–1220.

Al-Gazali, L. I. (2005). Attitudes toward genetic counseling in the United Arab Emirates. *Community Genetics, 8*(1), 48–51.

Amenson, C. S. and Lewinsohn, P. M. (1981). An investigation into the observed sex difference in prevalence of unipolar depression. *Journal of Abnormal Psychology, 90*(1), 1–13.

American Board of Genetic Counseling. (2008). *Graduate Programs in Genetic Counseling: Competencies.* ABGC Web site: www.abgc.net

Ani, C., Grantham-McGregor, S. and Muller, D. (2000). Nutritional supplementation in Down syndrome: Theoretical considerations and current status. *Developmental Medicine and Child Neurology, 42*(3), 207–213.

Anonymous (1995). Genome research risks abuse, panel warns. *Nature 378*(6557), 529.

Anonymous (1996). Statement: The Bell Curve. *Journal of Medical Ethics, 22*(3), 190.

Appels, M. C., Sitskoorn, M. M., Vollema, M. G. and Kahn, R. S. (2004). Elevated levels of schizotypal features in parents of patients with a family history of schizophrenia spectrum disorders. *Schizophrenia Bulletin, 30*(4), 781–790.

Bailey, A., Le Couteur, A., Gottesman, I., Bolton, P. et al. (1995). Autism as a strongly genetic disorder: Evidence from a British twin study. *Psychological Medicine, 25*(1), 63–77.

Barlow, D. P. (1993). Methylation and imprinting: From host defense to gene regulation? *Science, 260*(5106), 309–310.

Barrett, J. E. (2002). The emergence of behavioral pharmacology. *Molecular Interventions, 2*(8), 470–475.

Bates, G. and Lehrach, H. (1994). Trinucleotide repeat expansions and human genetic disease. *BioEssays: News and Reviews in Molecular, Cellular and Developmental Biology, 16*(4), 277–284.

Beck, E., Burnet, K. L. and Vosper, J. (2006). Birth-order effects on facets of extraversion. *Personality and Individual Differences, 40*, 953–959.

Beckett, L., Yu, Q. and Long, A. N. (2005). *The impact of fragile X: Prevalence, numbers affected, and economic impact.* Paper presented at the National Fragile X Awareness Day Research Seminar, Sacramento, CA.

Ben Zion, I. Z., Tessler, R., Cohen, L., Lerer, E. et al. (2006). Polymorphisms in the dopamine D_4 receptor gene (DRD_4) contribute to individual differences in human sexual behavior: Desire, arousal and sexual function. *Molecular psychiatry, 11*(8), 782–786.

Benjamin, J., Li, L., Patterson, C., Greenberg, B. D. et al. (1996). Population and familial association between the D_4 dopamine receptor gene and measures of novelty seeking. *Nature Genetics, 12*(1), 81–84.

Bennett, R. L., Motulsky, A. G., Bittles, A., Hudgins, L. et al. (2002). Genetic counseling and screening of consanguineous couples and their offspring: Recommendations of the National Society of Genetic Counselors. *Journal of Genetic Counseling, 11*(2), 97–119.

Bennett, R. L., Steinhaus, K. A., Uhrich, S. B., O'Sullivan, C. K. et al. (1995). Recommendations for standardized human pedigree nomenclature. Pedigree Standardization Task Force of the National Society of Genetic Counselors. *American Journal of Human Genetics, 56*(3), 745–752.

Berenbaum, S. A. and Hines, M. (1992). Early androgens are related to childhood sex-typed toy preferences. *Psychological Science, 3*(3), 203–206.

Berenbaum, S. A. and Resnick, S. M. (1997). Early androgen effects on aggression in children and adults with congenital adrenal hyperplasia. *Psychoneuroendocrinology, 22*(7), 505–515.

Bienvenu, T., Carrié, A., de Roux, N., Vinet, M. C. et al. (2000). *MECP2* mutations account for most cases of typical forms of Rett syndrome. *Human Molecular Genetics, 9*(9), 1377–1384.

Bird, T. D. (2007). Early-Onset Familial Alzheimer Disease. Gene reviews. Retrieved June 13, 2008, from www.ncbi.nlm. nih.gov/bookshelf/br.fcgi?book=gene&part=alzheimer-early.

Blair, I. P., Chetcuti, A. F., Badenhop, R. F., Scimone, A. et al. (2006). Positional cloning, association analysis and expression studies provide convergent evidence that the cadherin gene *FAT* contains a bipolar disorder susceptibility allele. *Molecular Psychiatry, 11*(4), 372–383.

Blatt, G. J., Chen, J. C., Rosene, D. L., Volicer, L. and Galler, J. R. (1994). Prenatal protein malnutrition effects on the serotonergic system in the hippocampal formation: An immunocytochemical, ligand binding, and neurochemical study. *Brain Research Bulletin, 34*(5), 507–518.

Bogaert, A. F. (2004). Asexuality: Prevalence and associated factors in a national probability sample. *Journal of Sex Research, 41*(3), 279–287.

Bögels, S. M., van Oosten, A., Muris, P. and Smulders, D. (2001). Familial correlates of social anxiety in children and adolescents. *Behaviour Research and Therapy, 39*(3), 273–287.

Bojesen, A., Juul, S. and Gravholt, C. H. (2003). Prenatal and postnatal prevalence of Klinefelter syndrome: A national registry study. *Journal of Clinical Endocrinology and Metabolism, 88*(2), 622–626.

Borkenau, P., Riemann, R., Angleitner, A. and Spinath, F. M. (2001). Genetic and environmental influences on observed personality: Evidence from the German Observational Study of Adult Twins. *Journal of Personality and Social Psychology, 80*(4), 655–668.

Borroni, B., Grassi, M., Costanzi, C., Archetti, S., et al. (2006). APOE genotype and cholesterol levels in lewy body dementia and Alzheimer disease: investigating genotype-phenotype effect on disease risk. *The American Journal of Geriatric Psychiatry, 14*(12), 1022–31.

Brown, E. S., Varghese, F. P. and McEwen, B. S. (2004). Association of depression with medical illness: Does cortisol play a role? *Biological Psychiatry, 55*(1), 1–9.

Bruder C. E., Piotrowski, A., Gijsbers, A. A., Andersson, R. et al. (2008). Phenotypically concordant and discordant monozygotic twins display different DNA copy-number-variation profiles. *American Journal of Human Genetics, 82*(3), 763–771.

Cadoret, R. J. (1978). Evidence for genetic inheritance of primary affective disorder in adoptees. *American Journal of Psychiatry, 135*(4), 463–466.

Campagne, F. and Weinstein, H. (1999). Schematic representation of residue-based protein context-dependent data: An application to transmembrane proteins. *Journal of Molecular Graphics and Modelling, 17*(3–4), 207–213.

Camperio-Ciani, A., Corna, F., and Capiluppi, C. (2004). Evidence for maternally inherited factors favouring male homosexuality and promoting female fecundity. *Proceedings of the Royal Society of London* B: *Biological Sciences, 271*(1554), 2217–2221.

Caspi, A., Sugden, K., Moffitt, T. E., Taylor, A. et al. (2003). Influence of life stress on depression: Moderation by a polymorphism in the *5-HTT* gene. *Science, 301*(5631), 386–389.

Cervino, A. C., Li, G., Edwards, S., Zhu, J. et al. (2005). Integrating QTL and high-density SNP analyses in mice to identify Insig2 as a susceptibility gene for plasma cholesterol levels. *Genomics, 86*(5), 505–517.

Coleman, P., Federoff, H. and Kurlan, R. A. (2004). Focus on the synapse for neuroprotection in Alzheimer disease and other dementias. *Neurology, 63*(7), 1155–1162.

Comings, D. E., Wu, S., Rostamkhani, M., McGue, M. et al. (2003). Role of the cholinergic muscarinic 2 receptor (CHRM2) gene in cognition. *Molecular Psychiatry, 8* (1), 10–11.

Cooper, R. M. and Zubek, J. P. (1958). Effects of enriched and restricted early environments on the learning ability of bright and dull rats. *Canadian Journal of Psychology, 12*(3), 159–164.

Coutelle, C., Themis, M., Waddington, S., Gregory, L. et al. (2003). The hopes and fears of in utero gene therapy for genetic disease: A review. *Placenta, 24*(Suppl. B), S114–S121.

Crocq, M. A., Mant, R., Asherson, P., Williams, J. et al (1992). Association between schizophrenia and homozygosity at the dopamine D_3 receptor gene. *Journal of Medical Genetics, 29*(12), 858–860.

Cryan, J. F. and Holmes, A. (2005). The ascent of mouse: Advances in modeling human depression and anxiety. *Nature Reviews: Drug Discovery, 4*(9), 775–790.

Csaba, A. and Papp, Z. (2003). Ethical dimensions of genetic counseling. *Clinics in Perinatology, 30*(1), 81–93.

Dale, K. S. and Landers, D. M. (1999). Weight control in wrestling: Eating disorders or disordered eating? *Medicine and Science in Sports and Exercise, 31*(10), 1382–1389.

Dalton, F. (1869). *Hereditary Genius: An Inquiry into Its Laws and Consequences.* London: Macmillan.

Darwin, C. (1859). *The Origin of Species by Means of Natural Selection, or the Preservation of Favored Races in the Struggle for Life.* London: Murray.

Daw, E. W., Payami, H., Nemens, E. J., Nochlin, D. et al. (2000). The number of trait loci in late-onset Alzheimer disease. *American Journal of Human Genetics, 66*(1), 196–204.

DeFries, J. C. and Alarcon, M. (1996). Genetics of specific reading disability. *Mental Retardation and Developmental Disabilities Research Reviews, 2*, 39–47.

DeSousa, E. A., Albert, R. H. and Kalman, B. (2002). Cognitive impairments in multiple sclerosis: A review. *American Journal of Alzheimer's Disease and Other Dementias, 17*(1), 23–29.

Doyle, A. E., Faraone, S. V., DuPre, E. P. and Biederman, J. (2001). Separating attention deficit hyperactivity disorder and learning disabilities in girls: A familial risk analysis. *American Journal of Psychiatry, 158*(10), 1666–1672.

Drevets, W. C., Frank, E., Price, J. C., Kupfer, D. J. et al. (2000). Serotonin type-1A receptor imaging in depression. *Nuclear Medicine and Biology, 27*(5), 499–507.

Driscoll, D. A., Salvin, J., Sellinger, B., Budarf, M. L. et al. (1993). Prevalence of 22q11 microdeletions in DiGeorge and velocardiofacial syndromes: Implications for genetic counselling and prenatal diagnosis. *Journal of Medical Genetics, 30*(10), 813–817.

Durand, M., Berton, O., Aguerre, S., Edno, L. et al. (1999). Effects of repeated fluoxetine on anxiety-related behaviours, central serotonergic systems, and the corticotropic axis in SHR and WKY rats. *Neuropharmacology, 38*(6), 893–907.

Dutch-Belgian Fragile X Consortium (1994). Fmr1 knockout mice: A model to study fragile X mental retardation. The Dutch-Belgian Fragile X Consortium. *Cell, 78*(1), 23–33.

Eastwood, H., Brown, K. M., Markovic, D. and Pieri, L. F. (2002). Variation in the *ESR1* and *ESR2* genes and genetic susceptibility to anorexia nervosa. *Molecular Psychiatry, 7*(1), 86–89.

Ebstein, R. P., Novick, O., Umansky, R., Priel, B. et al. (1996). Dopamine D_4 receptor (D_4DR) exon III polymorphism associated with the human personality trait of novelty seeking. *Nature Genetics, 12*(1), 78–80.

Edelman, G. (1987). *Neural Darwinism: The Theory of Neuronal Group Selection.* New York: Basic Books.

Eldredge, N. and Gould, S. J. (1972). Punctuated equilibria: An alternative to phyletic gradualism. In T. J. M. Schopf (Ed.), *Models in Paleobiology* (pp. 82–115). San Francisco: Freeman, Cooper.

Eley, T. C., Lichtenstein, P. and Moffitt, T. E. (2003). A longitudinal behavioral genetic analysis of the etiology of aggressive and nonaggressive antisocial behavior. *Development and Psychopathology, 15*(2), 383–402.

Emmons, S. W. and Lipton J (2003). Genetic basis of male sexual behavior. *Journal of Neurobiology, 54*(1), 93–110.

Faraone, S. V., Biederman, J., Lehman, B. K., Keenan, K. et al. (1993). Evidence for the independent familial transmission of attention deficit hyperactivity disorder and learning disabilities: Results from a family genetic study. *American Journal of Psychiatry, 150*(6), 891–895.

Farrer, L. A., Cupples, L. A., Haines, J. L., Hyman, B. et al. (1997). Effects of age, sex, and ethnicity on the association between apolipoprotein E genotype and Alzheimer disease: A meta-analysis. APOE and Alzheimer Disease Meta Analysis Consortium. *Journal of the American Medical Association, 278*(16), 1349–1356.

Fernández-Teruel, A., Escorihuela, R. M., Gray, J. A., Aguilar, R. et al. (2002). A quantitative trait locus influencing anxiety in the laboratory rat. *Genome Research, 12*(4), 618–626.

Fighting Autism (2008). Autism Rates. Fighting Autism Web site: www.fightingautism.org

Finn, C. T. and Smoller, J. W. (2006). Genetic counseling in psychiatry. *Harvard Review of Psychiatry, 14*(2), 109–121.

Fisher, S. E. and DeFries, J. C. (2002). Developmental dyslexia: Genetic dissection of a complex cognitive trait. *Nature Reviews: Neuroscience, 3*(10), 767–780.

Folstein, S. and Rutter, M. (1977). Genetic influences and infantile autism. *Nature, 265*(5596), 726–728.

Friedman, J. H., Trieschmann, M. E., Myers, R. H. and Fernandez, H. H. (2005). Monozygotic twins discordant for Huntington disease after 7 years. *Archives of Neurology, 62*(6), 995–997.

Friedman, J. M. (1999). Epidemiology of neurofibromatosis type 1. *American Journal of Medical Genetics, 89*(1), 1–6.

Friedmann, T. (1992). A brief history of gene therapy. *Nature Genetics, 2*(2), 93–98.

Fryns, J. P. (1984). The fragile X syndrome: A study of 83 families. *Clinical Genetics, 26*(6), 497–528.

Gallacher, J. E., Sweetnam, P. M., Yarnell, J. W., Elwood, P. C. et al. (2003). Is type A behavior really a trigger for coronary heart disease events? *Psychosomatic Medicine, 65*(3), 339–346.

Gaspar, H. B., Parsley, K. L., Howe, S., King, D. et al. (2004). Gene therapy of X-linked severe combined immunodeficiency by use of a pseudotyped gammaretroviral vector. *Lancet, 364*(9452), 2181–2187.

Gennarelli, M., Novelli, G., Andreasi Bassi, F., Martorell, L. et al. (1996). Prediction of myotonic dystrophy clinical severity based on the number of intragenic [CTG]n trinucleotide repeats. *American Journal of Medical Genetics, 65*(4), 342–347.

Genome research risks abuse, panel warns. (1995). *Nature, 378*(6557), 529.

Gerra, G., Garofano, L., Pellegrini, C., Bosari, S. et al. (2005). Allelic association of a dopamine transporter gene polymorphism with antisocial behaviour in heroin-dependent patients. *Addiction Biology, 10*(3), 275–281.

Glahn, D. C., Bearden, C. E., Niendam, T. A. and Escamilla, M. A. (2004). The feasibility of neuropsychological endophenotypes in the search for genes associated with bipolar affective disorder. *Bipolar Disorders, 6*(3), 171–182.

Goldblatt, M. J. and Schatzberg, A. F. (1991). Does treatment with antidepressant medication increase suicidal behavior? *International Clinical Psychopharmacology, 6*(4), 219–226.

Gosling, S. D. (2001). From mice to men: What can we learn about personality from animal research? *Psychological Bulletin, 127*(1), 45–86.

Gottesman, I. (1991). *Schizophrenia Genesis: The Origins of Madness*. New York: W. H. Freeman.

Gottesman, I. and Shields, J. (1982). *Schizophrenia: The Epigenetic Puzzle*. Cambridge University Press, Cambridge.

Grant, B. F., Hasin, D. S., Stinson, F. S., Dawson, et al. (2004). Prevalence, correlates, and disability of personality disorders in the United States: results from the national epidemiologic survey on alcohol and related conditions. *The Journal of Clinical Psychiatry, 65*(7), 948–58.

Gravholt, C. H. (2008). Epidemiology of Turner syndrome. *Lancet Oncology, 9*(3), 193–195.

Griffin, D. K., Abruzzo, M. A., Millie, E. A., Sheean, L. A. et al. (1995). Non-disjunction in human sperm: Evidence for an effect of increasing paternal age. *Human Molecular Genetics, 4*(12), 2227–2232.

Gringras, P. and Chen, W. (2001). Mechanisms for differences in monozygotic twins. *Early Human Development, 64*(2), 105–17.

Gruzelier, J. H. (2002). A review of the impact of hypnosis, relaxation, guided imagery and individual differences on aspects of immunity and health. *Stress, 5*(2), 147–163.

Gurevich, E. V., Bordelon, Y., Shapiro, R. M., Arnold, S. E. et al. (1997). Mesolimbic dopamine D_3 receptors and use of antipsychotics in patients with schizophrenia: A postmortem study. *Archives of General Psychiatry, 54*(3), 225–232.

Gusella, J. F., Wexler, N. S., Conneally, P. M., Naylor, S. L. et al. (1983). A polymorphic DNA marker genetically linked to Huntington's disease. *Nature, 306*(5940), 234–238.

Haas, D. W., Ribaudo, H. J., Kim, R. B., Tierney, C. et. al. (2004). Pharmacogenetics of efavirenz and central nervous system side effects: An Adult AIDS Clinical Trials Group study. *AIDS* (London), *18*(18), 2391–2400.

Hachey, D. L. and Chaurand, P. (2004). Proteomics in reproductive medicine: The technology for separation and identification of proteins. *Journal of Reproductive Immunology, 63*(1), 61–73.

Hadchouel, M., Farza, H., Simon, D., Tiollais, P. and Pourcel, C. (1987). Maternal inhibition of hepatitis B surface antigen gene expression in transgenic mice correlates with de novo methylation. *Nature, 329*(6138), 454–456.

Häfner, H. (2003). Gender differences in schizophrenia. *Psychoneuroendocrinology, 28*(Suppl. 2), 17–54.

Hajós, M., Fleishaker, J. C., Filipiak-Reisner, J. K., Brown, M. T. et al. (2004). The selective norepinephrine reuptake inhibitor antidepressant reboxetine: Pharmacological and clinical profile. *CNS Drug Reviews, 10*(1), 23–44.

Hall, J. G. (1990). Genomic imprinting: Review and relevance to human diseases. *American Journal of Human Genetics, 46*(5), 857–873.

Hamet, P. and Tremblay, J. (2005). Genetics and genomics of depression. *Metabolism: Clinical and Experimental, 54*(5) (Suppl. 1), 10–15.

Hammadeh, M. E., Sykoutris, A. and Schmidt, W. (2005). Relationship between body mass index and reproduction. *Current Women's Health Reviews, 1*(2), 131–142.

Han, C., McGue, M. K. and Iacono, W. G. (1999). Lifetime tobacco, alcohol and other substance use in adolescent Minnesota twins: Univariate and multivariate behavioral genetic analyses. *Addiction, 94*(7), 981–993.

Harder, T., Kube, E. and Gerke, V. (1992). Cloning and characterization of the human gene encoding p11: Structural similarity to other members of the S-100 gene family. *Gene, 113*(2), 269–274.

Hart, B. L. and Hart, L. A. (1985). Selecting pet dogs on the basis of cluster analysis of breed behavior profiles and gender. *Journal of the American Veterinary Medical Association, 186*(11), 1181–1185.

Hart, B. L. and Miller, M. F. (1985). Behavioral profiles of dog breeds. *Journal of the American Veterinary Medical Association, 186*(11), 1175–1180.

Healthtouch. (2006) *Dietary Restrictions for Infants and Children with PKU.* Healthtouch Web site: www.healthtouch.com accessed October 1, 2007.

Heimler, A. (1997). An oral history of the National Society of Genetic Counselors. *Journal of Genetic Counseling, 6*(3), 315–336.

Heron, W. T. (1935). The inheritance of maze learning ability in rats. *Journal of Comparative Psychology, 19,* 77–89.

Herrnstein, R. J. and Murray, C. (1994). *The Bell Curve: Intelligence and Class Structure in American Life.* New York: Free Press.

Hirschhorn, J. N., Lindgren, C. M., Daly, M. J., Kirby, A. et al. (2001). Genome-wide linkage analysis of stature in multiple populations reveals several regions with evidence of linkage to adult height. *American Journal of Human Genetics, 69*(1), 106–116.

Holliday, R. (1990). Genomic imprinting and allelic exclusion. *Development* (Suppl.), 125–129.

Hook, E. B., Cross, P. K. and Schreinemachers, D. M. (1983). Chromosomal abnormality rates at amniocentesis and in live-born infants. *Journal of the American Medical Association, 249*(15), 2034–2038.

Hopfer, C. J., Crowley, T. J. and Hewitt, J. K. (2003). Review of twin and adoption studies of adolescent substance use. *Journal of the American Academy of Child and Adolescent Psychiatry, 42*(6), 710–719.

Hovatta, I., Tennant, R. S., Helton, R., Marr, R. A. et al. (2005). Glyoxalase 1 and glutathione reductase 1 regulate anxiety in mice. *Nature, 438*(7068), 662–666.

Huang, Y. Y., Oquendo, M. A., Friedman, J. M., Greenhill, L. L. et al. (2003). Substance abuse disorder and major depression are associated with the human 5-HT1B receptor gene (HTR1B) G861C polymorphism. *Neuropsychopharmacology*: official publication of the American College of Neuropsychopharmacology, *28*(1), 163–169.

Hurd, Y. L. (2006). Perspectives on current directions in the neurobiology of addiction disorders relevant to genetic risk factors. *CNS Spectrums, 11*(11), 855–863.

International Human Genome Sequencing Consortium (2004). Finishing the euchromatic sequence of the human genome. *Nature, 431*(7011), 931–45.

Jääskeläinen, J., Mongan, N. P., Harland, S., and Hughes, I. A. (2006). Five novel androgen receptor gene mutations associated with complete androgen insensitivity syndrome. *Human mutation, 27*(3), 291.

Jang, K. L., Livesley, W. J. and Vernon, P. A. (1995). Alcohol and drug problems: A multivariate behavioural genetic analysis of co-morbidity. *Addiction, 90*(9), 1213–1221.

Jensen, A. R. (1980). Bias in Mental Testing. New York: Free Press.

Jensen, A. R. (1998). *The g Factor: The Science of Mental Ability.* Westport, CT: Praeger.

Johannessen, L., Strudsholm, U., Foldager, L. and Munk-Jørgensen, P. (2006). Increased risk of hypertension in patients with bipolar disorder and patients with anxiety compared to background population and patients with schizophrenia. *Journal of Affective Disorders, 95*(1–3), 13–17.

Johnson, S. W. and North, R. A. (1992). Opioids excite dopamine neurons by hyperpolarization of local interneurons. *Journal of Neuroscience, 12*(2), 483–488.

Johnston, N. (2005). And then there was Y: How the afterthought to the Human Genome Project broke the chromosomal mold. *Scientist, 19*, 24–25.

Jonas, B. S., Brody, D., Roper, M. and Narrow, W. E. (2003). Prevalence of mood disorders in a national sample of young American adults. *Social Psychiatry and Psychiatric Epidemiology, 38*(11), 618–624.

Joseph, J. (2001). Separated twins and the genetics of personality differences: A critique. *American Journal of Psychology, 114*(1), 1–30.

Kaye, W. H., Devlin, B., Barbarich, N., Bulik, C. M. et al. (2004). Genetic analysis of bulimia nervosa: Methods and sample description. *International Journal of Eating Disorders, 35*(4), 556–570.

Kendler, K. S., MacLean, C., Neale, M., Kessler, R. et al. (1991). The genetic epidemiology of bulimia nervosa. *American Journal of Psychiatry, 148*(12), 1627–1637.

Kendler, K. S., McGuire, M., Gruenberg, A. M., O'Hare, A. et al. (1993). The Roscommon Family Study. III. Schizophrenia-related personality disorders in relatives. *Archives of General Psychiatry, 50*(10), 781–788.

Kessler, R. C., Berglund, P., Demler, O., Jin, R. et al. (2005). Lifetime prevalence and age-of-onset distributions of DSM-IV disorders in the National Comorbidity Survey Replication. *Archives of General Psychiatry, 62*(6), 593–602.

Kirov, G., O'Donovan, M. C. and Owen, M. J. (2005). Finding schizophrenia genes. *Journal of Clinical Investigation, 115*(6), 1440–1448.

Klug, A. (1968). Rosalind Franklin and the discovery of the structure of DNA. *Nature, 219,* 808–810.

Klump, K. L. and Gobrogge, K. L. (2005). A review and primer of molecular genetic studies of anorexia nervosa. *International Journal of Eating Disorders, 37*(Suppl.), S43–S48, discussion S87–S89.

Konvicka, K., Campagne, F. and Weinstein, H. (2000). Interactive construction of residue-based diagrams of proteins: The RbDe Web service. *Protein Engineering, 13*(6), 395–396.

Koutsilieri, E., Rethwilm, A. and Scheller, C. (2007). The therapeutic potential of siRNA in gene therapy of neurodegenerative disorders. *Journal of Neural Transmission Supplementum* (72), 43–49.

Kremer, B., Goldberg, P., Andrew, S. E., Theilmann, J. et al. (1994). A worldwide study of the Huntington's disease mutation. The sensitivity and specificity of measuring CAG repeats. *The New England Journal of Medicine, 330*(20), 1401–1406.

Kringlen, E. (1991). Adoption studies in functional psychosis. *European Archives of Psychiatry and Clinical Neuroscience, 240*(6), 307–313.

Kuloglu, M., Atmaca, M., Tezcan, E., Ustundag, B. et al. (2002). Antioxidant enzyme and malondialdehyde levels in patients with panic disorder. *Neuropsychobiology, 46*(4), 186–189.

LaBuda, M. C., Gottesman, I. I. and Pauls, D. L. (1993). Usefulness of twin studies for exploring the etiology of childhood and adolescent psychiatric disorders. *American Journal of Medical Genetics, 48*(1), 47–59.

Ladwig, K. H., Marten-Mittag, B., Baumert, J., Löwel, H. and Döring, A. (2004). Case-finding for depressive and exhausted mood in the general population: Reliability and validity of a symptom-driven diagnostic scale. Results from the prospective MONICA/KORA Augsburg Study. *Annals of Epidemiology, 14*(5), 332–338.

Lakin, J. A., Steen, S. N. and Oppliger, R. A. (1990). Eating behaviors, weight loss methods, and nutrition practices among high school wrestlers. *Journal of Community Health Nursing, 7*(4), 223–234.

Lapidge, K. L., Oldroyd, B. P. and Spivak, M. (2002). Seven suggestive quantitative trait loci influence hygienic behavior of honey bees. *Die Naturwissenschaften, 89*(12), 565–568.

Laties, V. G. (1979). I. V. Zavadskii and the beginnings of behavioral pharmacology: An historical note and translation. *Journal of Experimental Analysis of Behavior, 32*(3), 463–472.

Leana-Cox, J., Pangkanon, S., Eanet, K. R., Curtin, M. S. et al. (1996). Familial DiGeorge/velocardiofacial syndrome with deletions of chromosome area 22q11.2: Report of five families with a review of the literature. *American Journal of Medical Genetics, 65*(4), 309–316.

Lesch, K. P. (2001). Molecular foundation of anxiety disorders. *Journal of Neural Transmission, 108*(6), 717–746.

Lieberman, J. A., Safferman, A. Z., Pollack, S., Szymanski, S. et al. (1994). Clinical effects of clozapine in chronic schizophrenia: Response to treatment and predictors of outcome. *American Journal of Psychiatry, 151*(12), 1744–1752.

Löwe, B., Gräfe, K., Ufer, C., Kroenke, K. et al. (2004). Anxiety and depression in patients with pulmonary hypertension. *Psychosomatic Medicine, 66*(6), 831–836.

Lynch, M. and Walsh, B. (1998). *Genetics and Analysis for Quantitative Traits.* Sunderland, MA: Sinauer Associates.

Lyons, M. J., Koenen, K. C., Buchting, F., Meyer, J. M. et al. (2004). A twin study of sexual behavior in men. *Archives of Sexual Behavior, 33*(2), 129–136.

MacQueen, G. M., Hajek, T. and Alda, M. (2005). The phenotypes of bipolar disorder: Relevance for genetic investigations. *Molecular Psychiatry, 10*(9), 811–826.

Majumdar, A. and Bose, S. K. (1968). DNA mediated genetic transformation of a human cancerous cell line cultured in vitro. *British Journal of Cancer, 22*(3), 603–613.

Mancama, D., Arranz, M. J. and Kerwin, R. W. (2002). Genetic predictors of therapeutic response to clozapine: Current status of research. *CNS Drugs, 16*(5), 317–324.

Mancama, D., Arranz, M. J. and Kerwin, R. W. (2003). Pharmacogenomics of psychiatric drug treatment. *Current Opinion in Molecular Therapeutics, 5*(6), 642–649.

Marshall, E. (2003). Gene therapy. Second child in French trial is found to have leukemia. *Science, 299*(5605), 320.

Mathews, C. A. and Reus, V. I. (2001). Assortative mating in the affective disorders: A systematic review and meta-analysis. *Comprehensive Psychiatry, 42*(4), 257–262.

Mathews, T. J. and Hamilton, B. E. (2002). Mean age of mother, 1970–2000. *National Vital Statistics Reports, 51*(1), 1–13.

McGrath, L. M., Smith, S. D. and Pennington, B. F. (2006). Breakthroughs in the search for dyslexia candidate genes. *Trends in Molecular Medicine, 12*(7), 333–341.

Mehler, P. S., Gray, M. C. and Schulte, M. (1997). Medical complications of anorexia nervosa. *Journal of Women's Health, 6*(5), 533–541.

Menza, M., Marin, H. and Opper, R. S. (2003). Residual symptoms in depression: Can treatment be symptom-specific? *Journal of Clinical Psychiatry, 64*(5), 516–523.

Merikangas, K. R. (1982). Assortative mating for psychiatric disorders and psychological traits. *Archives of General Psychiatry, 39*(10), 1173–1180.

Merikangas, K. R., Stolar, M., Stevens, D. E., Goulet, J. et al. (1998). Familial transmission of substance use disorders. *Archives of General Psychiatry, 55*(11), 973–979.

Miller, E. M. (1994). Intelligence and brain myelination: A hypothesis. *Personality and Individual Differences, 17*(6), 803–832.

Modrek, B. and Lee, C. (2002). A genomic view of alternative splicing. *Nature Genetics, 30*(1), 13–19.

Mortensen, P. B., Pedersen, C. B., Westergaard, T. and Wohlfahrt, J. (1999). Effects of family history and place and season of birth on the risk of schizophrenia. *New England Journal of Medicine, 340*(8), 603–608.

Moser, H. W. (2006) Therapy of X-linked adrenoleukodystrophy. *Neurotherapeutics, 3*(2), 246–253.

Muhle, R., Trentacoste, S. V. and Rapin, I. (2004). The genetics of autism. *Pediatrics, 113*(5), 472–486.

Muris, P., Merckelbach, H., Schmidt, H., Gadet, B. B. and Bogie, N. (2001). Anxiety and depression as correlates of self-reported behavioural inhibition in normal adolescents. *Behaviour Research and Therapy, 39*(9), 1051–1061.

Muris, P., Merckelbach, H., Wessel, I. and van de Ven, M. (1999). Psychopathological correlates of self-reported behavioural inhibition in normal children. *Behaviour Research and Therapy, 37*(6), 575–584.

Murphy, F. C., Rubinsztein, J. S., Michael, A., Rogers, R. D. et al. (2001). Decision-making cognition in mania and depression. *Psychological Medicine, 31*(4), 679–693.

Nestadt, G., Samuels, J., Riddle, M., Bienvenu, O. J. III et al. (2000). A family study of obsessive-compulsive disorder. *Archives of General Psychiatry, 57*(4), 358–363.

Nicholls, D. P. and Anderson, D. C. (1982). Clinical aspects of androgen deficiency in men. *Andrologia, 14*(5), 379–388.

Nielsen, T. O. (1997). Human germline gene therapy. *McGill Journal of Medicine 3*(2), 126–132.

Niimi, Y., Inoue-Murayama, M., Murayama, Y., Ito, S. and Iwasaki, T. (1999). Allelic variation of the D_4 dopamine receptor polymorphic region in two dog breeds, golden retriever and Shiba. *Journal of Veterinary Medical Science/Japanese Society of Veterinary Science, 61*(12), 1281–1286.

Noble, E. P. (2000). Addiction and its reward process through polymorphisms of the D_2 dopamine receptor gene: A review. *European Psychiatry, 15*(2), 79–89.

Nokelainen, P. and Flint, J. (2002). Genetic effects on human cognition: Lessons from the study of mental retardation syndromes. *Journal of Neurology, Neurosurgery, and Psychiatry, 72*(3), 287–296.

Nomura, T., Nakajima, H., Ryo, H., Li, L. Y. et al. (2004). Transgenerational transmission of radiation- and chemically induced tumors and congenital anomalies in mice: Studies of their possible relationship to induced chromosomal and molecular changes. *Cytogenetic and Genome Research, 104*(1–4), 252–260.

Noonan, C. W., Kathman, S. J. and White, M. C. (2002). Prevalence estimates for MS in the United States and evidence of an increasing trend for women. *Neurology, 58*(1), 136–138.

Nopola-Hemmi, J., Myllyluoma, B., Haltia, T., Taipale, M. et al. (2001). A dominant gene for developmental dyslexia on chromosome 3. *Journal of Medical Genetics, 38*(10), 658–664.

Nuechterlein, K. H. (1986). Childhood precursors of adult schizophrenia. *Journal of Child Psychology and Psychiatry and Allied Disciplines, 27*(2), 133–144.

Numakawa, T., Yagasaki, Y., Ishimoto, T., Okada, T. et al. (2004). Evidence of novel neuronal functions of dysbindin, a susceptibility gene for schizophrenia. *Human Molecular Genetics, 13*(21), 2699–2708.

O'Donnell, D. M. and Zoghbi, H. Y. (1995). Trinucleotide repeat disorders in pediatrics. *Current Opinion in Pediatrics, 7*(6), 715–725.

O'Donnell, W. T. and Warren, S. T. (2002). A decade of molecular studies of fragile X syndrome. *Annual Review of Neuroscience, 25*, 315–38.

Olson, R. K. (2002). Dyslexia: Nature and nurture. *Dyslexia* (Chichester, England), *8*(3), 143–159.

Onstad, S., Skre, I., Torgersen, S. and Kringlen, E. (1991). Twin concordance for DSM-III-R schizophrenia. *Acta Psychiatrica Scandinavica, 83*(5), 395–401.

Ostlund, H., Keller, E. and Hurd, Y. L. (2003). Estrogen receptor gene expression in relation to neuropsychiatric disorders. *Annals of the New York Academy of Sciences, 1007*, 54–63.

Ozaki, N., Goldman, D., Kaye, W. H., Plotnicov, K. et al. (2003). Serotonin transporter missense mutation associated with a complex neuropsychiatric phenotype. *Molecular Psychiatry, 8*(11), 933–936.

Paulhus, D. L., Trapnell, P. D. and Chen, D. (1999). Birth order effects on personality and achievement within families. *Psychological Science, 10*(6), 482–488.

Paunio, T., Tuulio-Henriksson, A., Hiekkalinna, T., Perola, M. et al. (2004). Search for cognitive trait components of schizophrenia reveals a locus for verbal learning and memory on 4q and for visual working memory on 2q. *Human Molecular Genetics, 13*(16), 1693–1702.

Peciña, S. and Berridge, K. C. (2005). Hedonic hot spot in nucleus accumbens shell: Where do mu-opioids cause increased hedonic impact of sweetness? *Journal of Neuroscience, 25*(50), 11777–11786.

Perna, G., Caldirola, D., Arancio, C. and Bellodi, L. (1997). Panic attacks: A twin study. *Psychiatry Research, 66*(1), 69–71.

Persico, A. M. and Bourgeron, T. (2006). Searching for ways out of the autism maze: Genetic, epigenetic and environmental clues. *Trends in Neurosciences, 29*(7), 349–358.

Persons, D. A. (2003). Update on gene therapy for hemoglobin disorders. *Current Opinion in Molecular Therapeutics, 5*(5), 508–516.

Pérusse, L. and Bouchard, C. (2000). Gene-diet interactions in obesity. *American Journal of Clinical Nutrition, 72*(5) (Suppl.), 1285S–1290S.

Pérusse, L., Rankinen, T., Zuberi, A., Chagnon, Y. C. et al. (2005). The human obesity gene map: the 2004 update. *Obesity Research, 13*(3), 381–490.

Plomin, R. (2003). Genetics, genes, genomics and *g. Molecular Psychiatry, 8*(1), 1–5.

Plomin, R. and Spinath, F. M. (2004). Intelligence: Genetics, genes, and genomics. *Journal of Personality and Social Psychology, 86*(1), 112–129.

Plomin, R., DeFries, J. C. and Loehlin, J. C. (1977). Genotype–environment interaction and correlation in the analysis of human behavior. *Psychological Bulletin, 84*(2), 309–322.

Plomin, R., DeFries, J. C., McClearn, G. E. and McGuffin, P. (2001). *Behavioral Genetics.* (4th ed.) New York: Worth Publishers.

Plomin, R., McClearn, G. E., Smith, D. L., Vignetti, S. et al. (1994). DNA markers associated with high versus low IQ: The IQ Quantitative Trait Loci (QTL) Project. *Behavior Genetics, 24*(2),107–118.

Posthuma, D., de Geus, E. J. C. and Boomsma, D. I. (2003). Genetic contributions to anatomical, behavioral and neurophysiological indices of cognition. In R. Plomin, J. DeFries, I. Craid and P. McGuffin (Eds.), *Behavioral Genetics in the Postgenomic Era* (pp. 141–161). Washington, DC: APA Books.

Price, R. A. and Gottesman, I. I. (1991). Body fat in identical twins reared apart: Roles for genes and environment. *Behavior Genetics, 21*(1), 1–7.

Prut, L. and Belzung, C. (2003). The open field as a paradigm to measure the effects of drugs on anxiety-like behaviors: A review. *European Journal of Pharmacology, 463*(1–3), 3–33.

Rabotti, G. F. (1963). Incorporation of DNA into a mouse tumor in vivo and in vitro. *Experimental Cell Research, 31*, 562–565.

Rankinen, T., Zuberi, A., Chagnon, Y. C., Weisnagel, S. J. et al. (2006). The human obesity gene map: The 2005 update. *Obesity, 14*(4), 529–644.

Reik, W., Collick, A., Norris, M. L., Barton, S. C. et al. (1987). Genomic imprinting determines methylation of parental alleles in transgenic mice. *Nature, 328*(6127), 248–251.

Resta, R., Biesecker, B. B., Bennett, R. L., Blum, S. et al. (2006). A new definition of Genetic Counseling: National Society of Genetic Counselors' Task Force report. *Journal of Genetic Counseling, 15*(2), 77–83.

Rhee, S. H., Hewitt, J. K., Young, S. E., Corley, R. P. et al. (2003). Genetic and environmental influences on substance initiation, use, and problem use in adolescents. *Archives of General Psychiatry, 60*(12), 1256–1264.

Ridley, R. M., Frith, C. D., Farrer, L. A. and Conneally, P. M. (1991). Patterns of inheritance of the symptoms of Huntington's disease suggestive of an effect of genomic imprinting. *Journal of Medical Genetics, 28*(4), 224–231.

Roberts, C. D., Stough, L. D. and Parrish, L. H. (2002). The role of genetic counseling in the elective termination of pregnancies involving fetuses with disabilities. *Journal of Special Education, 36*(1), 48–55.

Roebuck, T. M., Mattson, S. N. and Riley, E. P. (1998). A review of the neuroanatomical findings in children with fetal alcohol syndrome or prenatal exposure to alcohol. *Alcoholism, Clinical and Experimental Research, 22*(2), 339–344.

Rothenbuhler, W. C. (1964). Behavior genetics of nest cleaning in honey bees. IV. Responses of F_1 and backcross generations to disease killed brood. *American Zoologist, 4*, 111–123.

Ryu, S. H., Lee, S. H., Lee, H. J., Cha, J. H. et al. (2004). Association between norepinephrine transporter gene polymorphism and major depression. *Neuropsychobiology, 49*(4), 174–177.

Saad, F. A. and Jauniaux, E. (2002). Recurrent early pregnancy loss and consanguinity. *Reproductive Biomedicine Online, 5*(2), 167–170.

Saadat, M., Ansari-Lari, M. and Farhud, D. D. (2004). Consanguineous marriage in Iran. *Annals of Human Biology, 31*(2), 263–269.

Sachs, B. D. (1982). Role of striated penile muscles in penile reflexes, copulation, and induction of pregnancy in the rat. *Journal of Reproduction and Fertility, 66*(2), 433–443.

Sachs, B. D. and Meisel, R. L. (1994). The physiology of male sexual behavior. In E. Knobil and J. Neill (Eds.), *The Physiology of Reproduction* (pp. 3–105). New York: Raven Press.

Salstrom, J. L. (2007). X-inactivation and the dynamic maintenance of gene silencing. *Molecular genetics and metabolism, 92*(1-2), 56–62.

Sapienza, C., Peterson, A. C., Rossant, J. and Balling, R. (1987). Degree of methylation of transgenes is dependent on gamete of origin. *Nature, 328*(6127), 251–254.

Sapolsky, R. M. (2003). Gene therapy for psychiatric disorders. *American Journal of Psychiatry, 160*(2), 208–220.

Sarter, M. and Parikh, V. (2005). Choline transporters, cholinergic transmission and cognition. *Nature Reviews: Neuroscience, 6*(1), 48–56.

Savolainen, P., Zhang, Y. P., Luo, J., Lundeberg, J. and Leitner, T. (2002). Genetic evidence for an East Asian origin of domestic dogs. *Science, 298*(5598), 1610–1613.

Sayre, A. (1975). *Rosalind Franklin and DNA.* New York: Norton.

Scapagnini, U. (1992). Psychoneuroendocrinoimmunology: The basis for a novel therapeutic approach in aging. *Psychoneuroendocrinology, 17*(4), 411–420.

Schiepers, O. J., Wichers, M. C. and Maes, M. (2005). Cytokines and major depression. *Progress in Neuro-Psychopharmacology and Biological Psychiatry, 29*(2), 201–217.

Serretti, A., Artioli, P. and Quartesan, R. (2005). Pharmacogenetics in the treatment of depression: Pharmacodynamic studies. *Pharmacogenetics and Genomics, 15*(2), 61–67.

Serretti, A., Lilli, R., Di Bella, D., Bertelli, S. et al. (1999). Dopamine receptor D_4 gene is not associated with major psychoses. *American Journal of Medical Genetics, 88*(5), 486–491.

Shaffer, H. J., LaPlante, D. A., LaBrie, R. A., Kidman, R. C. et al. (2004). Toward a syndrome model of addiction: Multiple expressions, common etiology. *Harvard Review of Psychiatry, 12*(6), 367–374.

Shatz, S. M. (2005). The psychometric properties of the behavioral inhibition scale in a college-aged sample. *Personality and Individual Differences, 39,* 331–339.

Simerly, R. B. (2002). Wired for reproduction: Organization and development of sexually dimorphic circuits in the mammalian forebrain. *Annual Review of Neuroscience, 25,* 507–536.

Singhrao, S. K., Thomas, P., Wood, J. D., MacMillan, J. C. et al. (1998). Huntingtin protein colocalizes with lesions of neurodegenerative diseases: An investigation in Huntington's, Alzheimer's, and Pick's diseases. *Experimental Neurology, 150*(2), 213–222.

Skaletsky, H., Kuroda-Kawaguchi, T., Minx, P. J., Cordum, H. S. et al. (2003). The male-specific region of the human Y chromosome is a mosaic of discrete sequence classes. *Nature, 423*(6942), 825–837.

Skinner, B. F. and Heron, W. T. (1937). Effects of caffeine and benzedrine upon conditioning and extinction. *Psychological Record, 1,* 340–346.

Smith, L. B., Sapers, B., Reus, V. I. and Freimer, N. B. (1996). Attitudes towards bipolar disorder and predictive genetic testing among patients and providers. *Journal of Medical Genetics, 33*(7), 544–549.

Smits, K. M., Smits, L. J., Schouten, J. S., Stelma, F. F. et al. (2004). Influence of SERTPR and STin2 in the serotonin transporter gene on the effect of selective serotonin reuptake inhibitors in depression: A systematic review. *Molecular Psychiatry, 9*(5), 433–441.

Sommer, V. and Vasey, P. L. (Eds.) (2006). *Homosexual Behaviour in Animals: An Evolutionary Perspective.* Cambridge: Cambridge University Press.

Souter, V. L., Parisi, M. A., Nyholt, D. R., Kapur, R. P. et al. (2007). A case of true hermaphroditism reveals an unusual mechanism of twinning. *Human Genetics, 121*(2), 179–185.

Spillantini, M. G., Murrell, J. R., Goedert, M., Farlow, M. R. et al. (1998). Mutation in the tau gene in familial multiple system tauopathy with presenile dementia. *Proceedings of the National Academy of Sciences of the United States of America, 95*(13), 7737–7741.

Spinath, F. M., Harlaar, N., Ronald, A. and Plomin, R. (2004). Substantial genetic influence on mild mental impairment in early childhood. *American Journal of Mental Retardation, 109*(1), 34–43.

Spivak, M. and Reuter, G. (2005) *A Sustainable Approach to Controlling Honey Bee Diseases and Varroa Mites: Breeding for Hygienic Behavior.* Sustainable Agriculture Research Education. SARE fact sheet # 03AGI2005. Retrieved June 25, 2008 from www.sare.org/publications/factsheet/0305_03.htm.

Statement: *The Bell Curve.* (1996). *Journal of Medical Ethics, 22*(3), 190.

Stefanis, N. C., Bresnick, J. N., Kerwin, R. W., Schofield, W. N. and McAllister, G. (1998). Elevation of D_4 dopamine receptor mRNA in postmortem schizophrenic brain. *Brain Research: Molecular Brain Research, 53*(1–2), 112–119.

Stein, L. D. (2004). Human genome: End of the beginning. *Nature, 431*(7011), 915–916.

Stout, M. (2006). *The Sociopath Next Door: The Ruthless* versus *the Rest of Us.* New York: Broadway Books..

Strober, M., Freeman, R., Lampert, C., Diamond, J. et al. (2000). Controlled family study of anorexia nervosa and bulimia nervosa: Evidence of shared liability and transmission of partial syndromes. *American Journal of Psychiatry, 157*(3), 393–401.

Strulovici, Y., Leopold, P. L., O'Connor, T. P., Pergolizzi, R. G. et al. (2007). Human embryonic stem cells and gene therapy. *Molecular Therapy, 15*(5), 850–866.

Sullivan, P. F., Neale, M. C. and Kendler, K. S. (2000). Genetic epidemiology of major depression: Review and meta-analysis. *American Journal of Psychiatry, 157*(10), 1552–1562.

Svenningsson, P., Chergui, K., Rachleff, I., Flajolet, M. et al. (2006). Alterations in 5-HT1B receptor function by p11 in depression-like states. *Science, 311*(5757), 77–80.

Swain, J. L., Stewart, T. A. and Leder, P. (1987). Parental legacy determines methylation and expression of an autosomal transgene: A molecular mechanism for parental imprinting. *Cell, 50*(5), 719–727.

Szatmari, P., Jones, M. B., Zwaigenbaum, L. and MacLean, J. E. (1998). Genetics of autism: Overview and new directions. *Journal of Autism and Developmental Disorders, 28*(5), 351–368.

Takahashi, N., Kaji, H., Yanagida, M., Hayano, T. et al. (2003). Proteomics: Advanced technology for the analysis of cellular function. *Journal of Nutrition, 133*(6) (Suppl. 1), 2090S–2096S.

Tang, Y. P., Shimizu, E., Dube, G. R., Rampon, C. et al. (1999). Genetic enhancement of learning and memory in mice. *Nature, 401*(6748), 63–69.

Tate, S. K. and Goldstein, D. B. (2004). Will tomorrow's medicines work for everyone? *Nature Genetics, 36*(11) (Suppl.), S34–S42.

Teicher, M. H., Glod, C. A. and Cole, J. O. (1993). Antidepressant drugs and the emergence of suicidal tendencies. *Drug Safety: An International Journal of Medical Toxicology and Drug Experience, 8*(3), 186–212.

Thaker, G., Adami, H., Moran, M., Lahti, A. et al. (1993). Psychiatric illnesses in families of subjects with schizophrenia-spectrum personality disorders: High morbidity risks for unspecified functional psychoses and schizophrenia. *American Journal of Psychiatry, 150*(1), 66–71.

Thapar, A., Langley, K., Fowler, T., Rice, F. et al. (2005). Catechol O-methyltransferase gene variant and birth weight predict early-onset antisocial behavior in children with attention-deficit/hyperactivity disorder. *Archives of General Psychiatry, 62*(11), 1275–1278.

Thompson, W. R. (1954). The inheritance and development of intelligence. *Proceedings of the Association for Research in Nervous and Mental Disease, 33,* 209–231.

Toga, A. W. and Thompson, P. M. (2005). Genetics of brain structure and intelligence. *Annual Review of Neuroscience, 28,* 1–23.

Trottier, Y., Biancalana, V. and Mandel, J. L. (1994). Instability of CAG repeats in Huntington's disease: relation to parental transmission and age of onset. *Journal of Medical Genetics, 31*(5), 377–382.

Tryon, R. C. (1940). Genetic differences in maze learning ability in rats. *Yearbook of the National Society for Studies in Education, 39,* 111–119.

Tsuang, M. (2000). Schizophrenia: Genes and environment. *Biological Psychiatry, 47*(3), 210–220.

van Oers, K., Drent, P. J., de Goede, P. and van Noordwijk, A. J. (2004). Realized heritability and repeatability of risk-taking behaviour in relation to avian personalities. *Proceedings of the Royal Society of London* B: *Biological Sciences, 271*(1534), 65–73.

Vaswani, M. and Kalra, H. (2004). Selective serotonin re-uptake inhibitors in anorexia nervosa. *Expert Opinion on Investigational Drugs, 13*(4), 349–357.

Venter, J. C., Adams, M. D., Myers, E. W., Li, P. W. et al. (2001). The sequence of the human genome. *Science, 291*(5507), 1304–1351.

Wang, Q. J., Lu, C. Y., Li, N., Rao, S. Q. et al. (2004). Y-linked inheritance of non-syndromic hearing impairment in a large Chinese family. *Journal of Medical Genetics, Electronic Letter, 41,*e80.

Wassef, A., Baker, J. and Kochan, L. D. (2003). GABA and schizophrenia: A review of basic science and clinical studies. *Journal of Clinical Psychopharmacology, 23*(6), 601–640.

Watson, J. D. and Crick, F. H. (1953). Molecular structure of nucleic acids; a structure for deoxyribose nucleic acid. *Nature; 171*(4356), 737–738.

Watson, N. F., Goldberg, J., Arguelles, L. and Buchwald, D. (2006). Genetic and environmental influences on insomnia, daytime sleepiness, and obesity in twins. *Sleep, 29*(5), 645–649.

Whitam, F. L., Diamond, M. and Martin, J. (1993). Homosexual orientation in twins: A report on 61 pairs and three triplet sets. *Archives of Sexual Behavior, 22*(3), 187–206.

Wirdefeldt, K., Gatz, M., Schalling, M. and Pedersen, N. L. (2004). No evidence for heritability of Parkinson disease in Swedish twins. *Neurology, 63*(2), 305–311.

Witkin, H. A., Mednick, S. A., Schulsinger, F., Bakke-strom, E. et al. (1976). Criminality in XYY and XXY men. *Science, 193*(4253), 547–555.

Wolff, P. H. and Melngailis, I. (1994). Family patterns of developmental dyslexia: Clinical findings. *American Journal of Medical Genetics, 54*(2), 122–131.

Wong, D. T., Bymaster, F. P. and Engleman, E. A. (1995). Prozac (fluoxetine, Lilly 110140), the first selective serotonin uptake inhibitor and an antidepressant drug: Twenty years since its first publication. *Life Sciences, 57*(5): 411–441.

Wood, F. B. and Grigorenko, E. L. (2001). Emerging issues in the genetics of dyslexia: A methodological preview. *Journal of Learning Disabilities, 34*(6), 503–511.

Wu, E. Q., Shi, L., Birnbaum, H., Hudson, T. et al. (2006). Annual prevalence of diagnosed schizophrenia in the USA: a claims data analysis approach. *Psychological Medicine, 36*(11), 1535–1540.

Yang, J., Feng, G., Zhang, J., Hui, Z. et al. (2001). Is *ApoE* gene a risk factor for vascular dementia in Han Chinese? *International Journal of Molecular Medicine, 7*(2), 217–219.

Yeargin-Allsopp, M., Rice, C., Karapurkar, T., Doernberg, N. et al. (2003). Prevalence of autism in a U.S. metropolitan area. *Journal of the American Medical Association, 289*(1), 49–55.

Zhang, Y., Proenca, R., Maffei, M., Barone, M. et al. (1994). Positional cloning of the mouse *obese* gene and its human homologue. *Nature, 372*(6505), 425–432.

Zlotogora, J. (2005). Is there an increased birth defect risk to children born to offspring of first cousin parents? *American Journal of Medical Genetics, 137*(3), 342.

Zubenko, G. S., Hughes, H. B. III, Maher, B. S., Stiffler, J. S. et al. (2002). Genetic linkage of region containing the *CREB1* gene to depressive disorders in women from families with recurrent, early-onset, major depression. *American Journal of Medical Genetics, 114*(8), 980–987.

Zuchner, S., Cuccaro, M. L., Tran-Viet, K. N., Cope, H. et al. (2006). SLITRK1 mutations in trichotillomania. *Molecular Psychiatry, 11*(10), 887–889.

INDEX

Numbers in *italic* indicate the information will be found in a figure or table.